Sustainable Civil Infrastructures

About this Series

Sustainable Infrastructure impacts our well-being and day-to-day lives. The infrastructures we are building today will shape our lives tomorrow. The complex and diverse nature of the impacts due to weather extremes on transportation and civil infrastructures can be seen in our roadways, bridges, and buildings. Extreme summer temperatures, droughts, flash floods, and rising numbers of freeze-thaw cycles pose challenges for civil infrastructure and can endanger public safety. We constantly hear how civil infrastructures need constant attention, preservation, and upgrading. Such improvements and developments would obviously benefit from our desired book series that provide sustainable engineering materials and designs. The economic impact is huge and much research has been conducted worldwide. The future holds many opportunities, not only for researchers in a given country, but also for the worldwide field engineers who apply and implement these technologies. We believe that no approach can succeed if it does not unite the efforts of various engineering disciplines from all over the world under one umbrella to offer a beacon of modern solutions to the global infrastructure. Experts from the various engineering disciplines around the globe will participate in this series, including: Geotechnical, Geological, Geoscience, Petroleum, Structural, Transportation, Bridge, Infrastructure, Energy, Architectural, Chemical and Materials, and other related Engineering disciplines.

More information about this series at http://www.springer.com/series/15140

Meng-Chia Weng · Jeffrey Lee · Yong Liu
Editors

Current Geotechnical Engineering Aspects of Civil Infrastructures

Proceedings of the 5th GeoChina International Conference 2018 – Civil Infrastructures Confronting Severe Weathers and Climate Changes: From Failure to Sustainability, held on July 23 to 25, 2018 in HangZhou, China

Springer

Editors
Meng-Chia Weng
Department of Civil Engineering
National Chiao Tung University
Hsinchu, Taiwan

Yong Liu
School of Water Resources
and Hydropower Engineering
Wuhan University
Wuhan, China

Jeffrey Lee
Australian Road Research Board (ARRB)
Brisbane, QLD, Australia

ISSN 2366-3405 ISSN 2366-3413 (electronic)
Sustainable Civil Infrastructures
ISBN 978-3-319-95749-4 ISBN 978-3-319-95750-0 (eBook)
https://doi.org/10.1007/978-3-319-95750-0

Library of Congress Control Number: 2018948648

This Springer imprint is published by the registered company Springer Nature Switzerland AG
The registered company address is: Gewerbestrasse 11, 6330 Cham, Switzerland

Contents

Numerical Analysis of Tunnel Lining and Ground Behaviour During Shield Tunneling—A Case Study in Japan 1
Grant Hasan, Alireza Afshani, and Akagi Hirokazu

Sustainable Reconstruction in Conditions of Dense Urban Development 13
Talal Awwad, Vladimir Gruzin, and Vladimir Kim

Influence of Unit Weight of Slurry-Deposited Coal Ash on Stability of Ash Dykes Raised by Upstream Method 24
Aali Pant, Manoj Datta, and Gunturi Venkata Ramana

Impact Assessment of Climate Change in Texas Pavements and Resiliency Strategy 38
Megha Sharma, Sundeep Inti, and Vivek Tandon

Implementation of Computational Contact Analysis in Geotechnical Engineering 54
Daqing Xu, Luis G. Vasquez, and William M. Isenhower

Performance of Geocell Reinforced Embankment over Soft Soil Deposit 69
Lalima Banerjee, Sowmiya Chawla, and Gupinath Bhandari

A Study on Utilization of Mine Overburden as a Replacement of Base and Sub-base Layers on Rural Roads 83
Brajendra Nath, Sowmiya Chawla, and Randhir Kumar Gupta

A New Equation for SWCCs of Unsaturated Soils 101
Suched Likitlersuang and Martin D. Liu

Effects of Dry Density and Water Content on Mechanical Properties of Sand-Bentonite Buffer Material 109
Janaka J. Kumara and Takeshi Kodaka

Geotechnical Investigation on Compacted Clay Liner Blended with Reactive Material 124
Polapala Sai Pradeep, Mayakrishnan Muthukumar, and Sanjay Kumar Shukla

Author Index 133

Introduction

With increasing urbanization rates and development of society, advancement in geotechnical technologies is essential to the construction of infrastructures. Geotechnical Investigation is the first step of applying scientific methods and engineering principles to obtain solutions of civil engineering problems. This volume brings together scientific experts in different areas that contribute to the state of the art in geotechnical engineering, such as characterization of geomaterials, slope stability, tunneling, sustainability in geohazards, and some other geotechnical issues that are quite relevant in today's world. This volume is a part of the Proceedings of the 5th GeoChina International Conference 2018—Civil Infrastructures Confronting Severe Weathers and Climate Changes: From Failure to Sustainability, HangZhou, China.

Numerical Analysis of Tunnel Lining and Ground Behaviour During Shield Tunneling—A Case Study in Japan

Grant Hasan(✉), Alireza Afshani, and Akagi Hirokazu

Soil and Geotechnical Environment, Waseda University, Tokyo, Japan
grant_hasan@hotmail.com, aafshani@aoni.waseda.jp,
akagi@waseda.jp

Abstract. Numerical analyses are widely used to perform a detailed analysis of the tunnel excavations. During the tunneling, parameters such as grouting, overburden, and lining property affect the magnitude and shape of ground surface settlement. By using the measurement data of a shield tunneling excavation case in Japan, the authors tried to investigate the effects of influencing parameters on the ground surface settlement. Furthermore, bending moment, axial force and vertical displacements of tunnel are calculated and compared with the field measurement data. Ground settlement trough is analyzed and compared with the other cases of tunneling in previous literatures. The numerical analyses are conducted in two cases of taking consideration of buoyancy force and without buoyancy force. The accuracy of the numerical analysis is checked by comparing the numerical results with the field measurement data.

1 Introduction

Numerical analyses of tunnels are commonly used to predict the ground and tunnel behavior during tunneling construction. Different models on numerical analysis are done to predict the ground movement during shield tunneling. The complexity of the model is different depending on the site condition such as, the ground condition and the construction method involved.

In this study, by using a 2D modeling of a tunneling excavation case, the tunnel lining behavior, ground volume loss, and ground trough shape are investigated numerically. In order to evaluate the effect of buoyancy in tunnel excavation, the analyses are done in two cases of with and without buoyancy force consideration. The ground volume loss calculated in this case is compared with other tunneling case studies with similar soil properties. Ground surface settlement trough is obtained and compared with Gaussian approximation curve in the two cases of with and without buoyancy. Ground normalized surface settlement curve is also compared with similar cases of tunneling induced surface settlement trough.

M.-C. Weng et al. (eds.), *Current Geotechnical Engineering Aspects of Civil Infrastructures*, Sustainable Civil Infrastructures,
https://doi.org/10.1007/978-3-319-95750-0_1

2 Case Study Review

The case study is a shield tunneling construction work between Kotake-Mukaihara station and Senkawa station of Tokyo Metro Yurakucho Lines. Figure 1 shows the top view map of the Yurakucho lines of A and B. The geological formation is mainly soft soil and clay mixture, the soil longitudinal profile and the soil layer properties are showed on Fig. 2 and Table 1 respectively. The average overburden in Tunnel A is 12.6 m which is used in the numerical analysis model. As shown on Fig. 3, the tunnel lining is an elliptical shape pre-cast concrete lining with the height of 6.6 m and width of 5.5 m, with segmental ring length of 1.5 m.

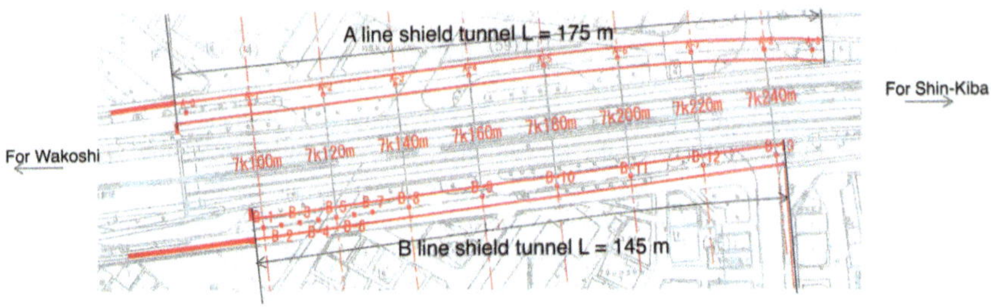

Fig. 1. Top view map of Yurakucho tunnel lines of A and B

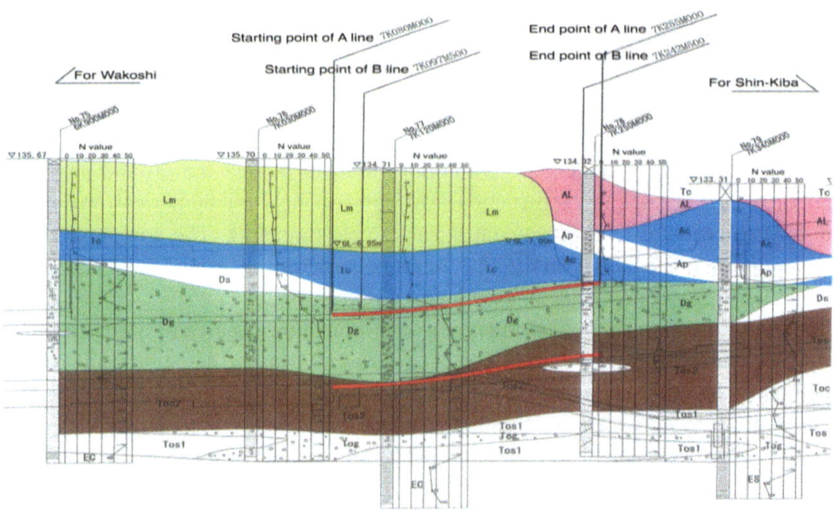

Fig. 2. Longitudinal profile of soil layers and tunnel lines

Table 1. Soil layer properties

Layer	Description	N_{SPT}-value	Unit weight, γ (kg/m^3)	Elastic modulus, E (kh/m^2)	Poison's ratio, ν	Cohesion, c (kN/m^2)	Internal friction angle, Φ (°)	Coefficient lateral earth pressure, K_0
AL	Very soft clay	0–2	Assumed to be same as Lm					
Ap	Cohesive clay	1–2						
Lm	Loam	4–6	1200	8×10^5	0.45	60	0	1
Lc	Soft clay	3–10	1300	8×10^5	0.45	90	0	1
Mg	Sand and gravel	20–50	2000	2.5×10^6	0.3	0	35	0.5
Tos2	Sand	30–50	1800	9.5×10^6	0.3	0	35	0.5
Tog	Gravel	Over 50	1800	9.5×10^6	0.3	0	35	0.5

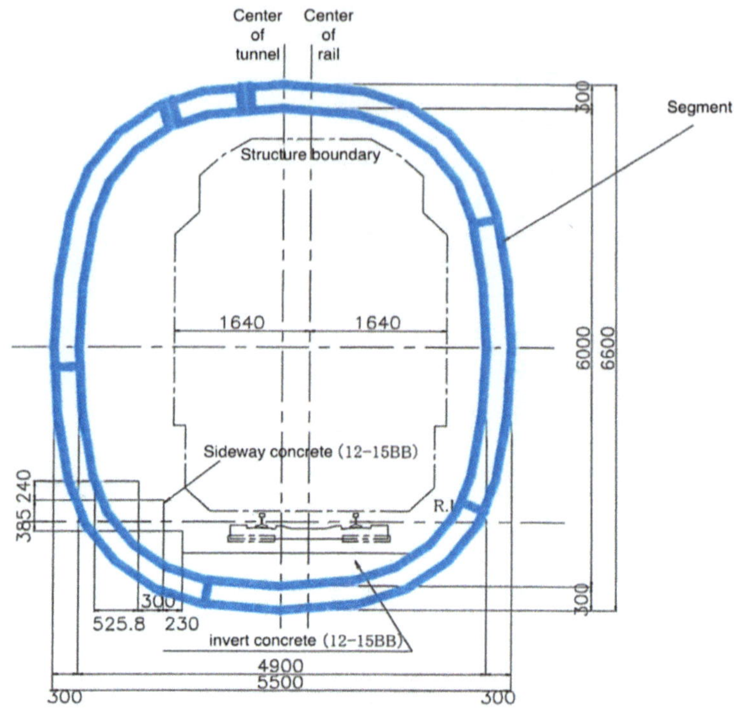

Fig. 3. Cross section of the tunnel

3 The Buoyancy Uplift Force Consideration

The buoyancy is an uplift force that is generated around the tunnel due to the difference in weight of the completed tunnel and the excavated soil. The buoyancy effect will only occur in the saturated soft soil condition (Verruijt and Strack 2008).

The uplift force from buoyancy can be calculated as follow:

$$F = (1 - w_t)\gamma A \tag{1}$$

which F represent the buoyancy uplift force (N), w_t is the ratio of the mass density of tunnel to the mass density of the soil, A is area of the tunnel (m^2) and γ is the mass density of soil (kN/m^3).

In order to evaluate the effect of buoyancy in tunneling construction, analyses are done in two cases of with and without buoyancy. The analyses are conducted using a finite element analysis program, Midas GTS. In the analyses, the buoyancy force is applied on the nodes around the tunnel lining.

4 Lining Analysis

In this section, the behaviour of the Yurakucho tunnel lining and ground surface settlement trough is demonstrated using the data of introduced case study.

4.1 Modeling Steps

In this part, numerical modeling steps are explained. The parameters listed in Table 2 are used in numerical analysis. The analyses consist of three stages, the which are defined in Table 3.

Table 2. Material properties used in the model

Soil type	Layer thickness (m)	Mass density, γ (kg/m^3)	Young's modulus, E (kg/m^2)	Poisson's ratio, ν	Cohesion, c (kN/m^2)	Internal friction angle, Φ (°)	Drainage condition
Lm	6.8	1200	8×10^5	0.45	60	0	Undrained
Lc	3.4	1300	8×10^5	0.45	90	0	Undrained
Mg	7.8	2000	2.5×10^6	0.3	0	35	Drained
Tos	4.8	1800	9.5×10^6	0.3	0	35	Drained
Tog	2.6	1800	9.5×10^6	0.3	0	35	Drained
Concrete	–	2400	3.82×10^9	0.15	–	–	–

Table 3. Construction steps details

Steps	Description
Initial	All the soil elements are set, the boundary condition and gravity load are applied to take the consideration of the soil self-weight
Excavation	The soil element inside the tunnel area are removed, 10% lining bending stiffness are applied to simulate that the soil elements are excavated slowly. The buoyancy force is applied in this step
Final	The 100% lining bending stiffness is applied

4.2 Vertical Displacement

Figure 4 shows the tunnel lining deformation with and without buoyancy consideration, the tunnel lining displacement on the crown for the case of with and without buoyancy are 10.66 and 13.64 mm respectively. The ground surface settlement and lining deformation on the last step of the excavation are shown in the Fig. 5. From these figures, it can be seen that the deformation decreases by buoyancy consideration. This is due to the uplift force caused by the excavation. The detail values of the vertical displacement are shown on Table 4. It lists the vertical displacement at ground and tunnel lining crown and invert for the two cases of with and without buoyancy consideration.

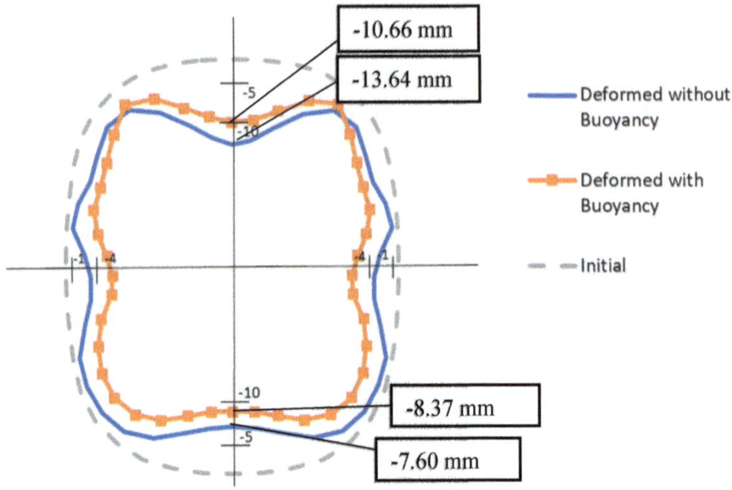

Fig. 4. Numerical results of lining deformation on the last step of construction

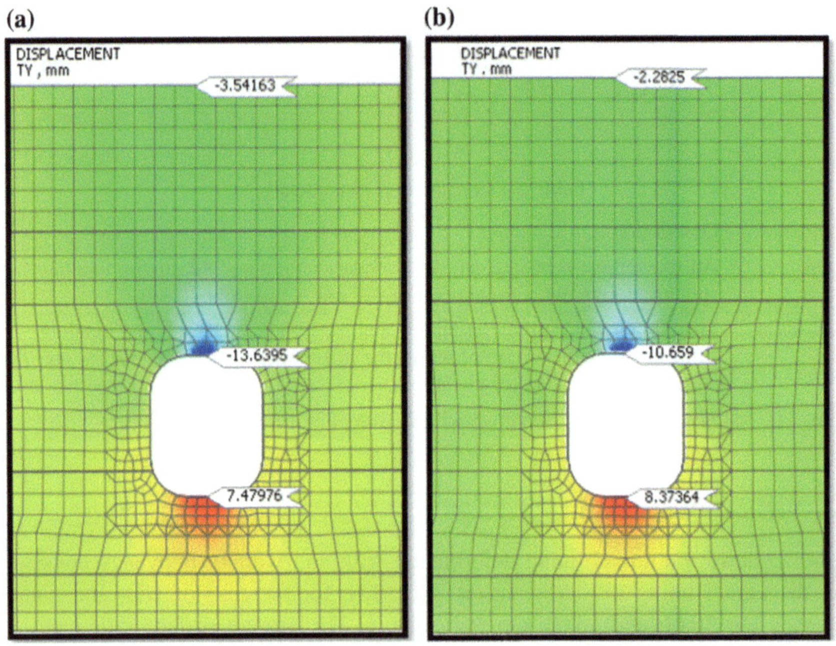

Fig. 5. Numerical results of ground surface settlement and lining deformation on the last step of construction for two cases of **a** without buoyancy and **b** with buoyancy

Table 4. Vertical displacement caused by tunneling

	δ_{WOB} (mm)	δ_{WB} (mm)	$\Delta\delta$ (mm)
Tunnel lining crown	−13.64	−10.66	3.08
Tunnel lining invert	−7.600	−8.373	1.66
Ground settlement	−3.82	−2.22	1.60

δ_{WOB}: Displacement without considering buoyancy force
δ_{WB}: Displacement with considering buoyancy force
$\Delta\delta$: Displacement difference in vertical displacement between the two cases of δ_{WB} and δ_{WOB}

Using Eq. (2), the Gaussian approximation are also depicted in Fig. 6. $S(y)$ is the settlement, y as the offset coordinate from tunnel center line, and i as the point of inflection. and S_{vmax} is the maximum ground surface settlement.

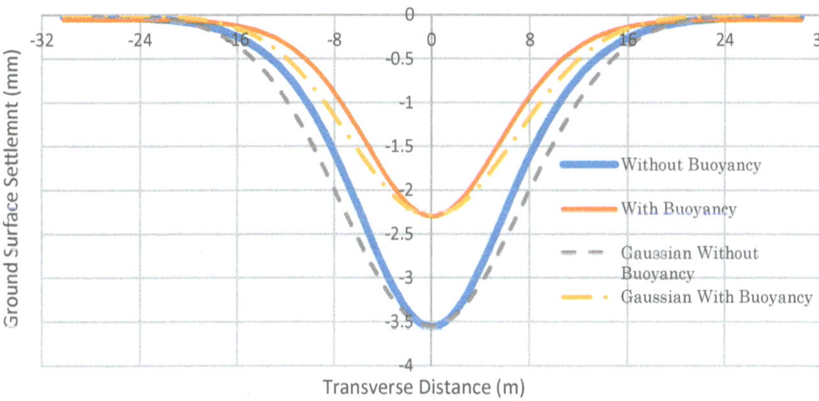

Fig. 6. Comparison of numerically obtained ground surface settlement with theoretical Gaussian formula for two cases of with and without buoyancy

$$S(y) = S_{vmax} \times e^{-\frac{y^2}{2i^2}} \tag{2}$$

The curves show that the Gaussian approximation has a wider trough compared to the Yurakucho field measurement. This is due to the existence of top soft soil layers in the Yurakucho site. The field ground surface settlement trough curves fit to the Gaussian approximation curve.

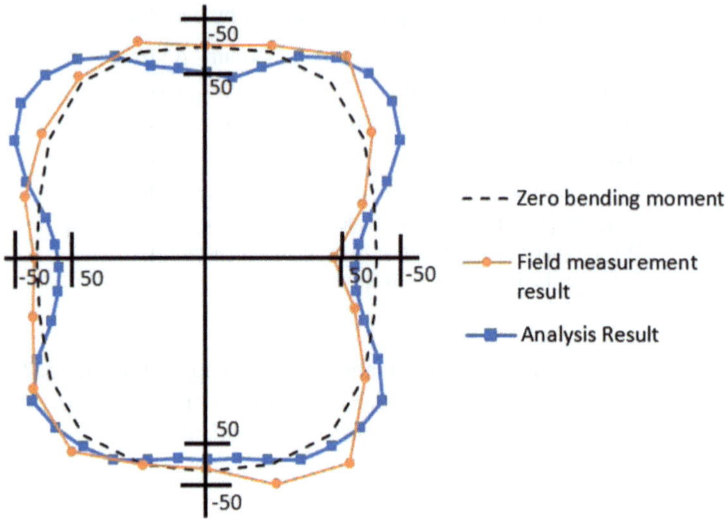

Fig. 7. Comparison of field bending moment data measurement to analysis measurement

4.3 Bending Moment and Axial Force

The numerical analysis results of the bending moment and axial force are shown in Figs. 7 and 8. The measurement data of Yurakucho tunnel are also shown in the same figure comparison. Numerical results have overall same behavior with measured field data. On Fig. 8, the maximum axial force is close to the measurement data.

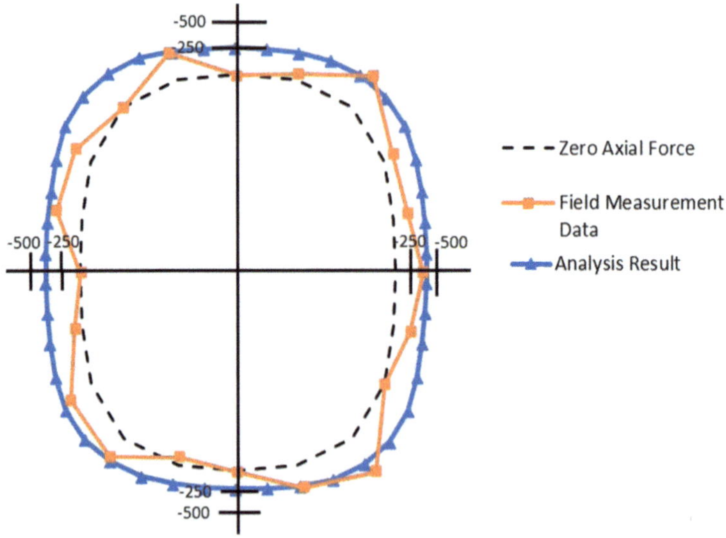

Fig. 8. Comparison of field axial force data measurement to analysis measurement

5 Ground Volume Loss

Volume loss is defined as the volume of ground loss proportion of final tunnel volume in a plane. Equations (3) and (4) present the theoretical method to calculate the volume loss.

$$A_s = \sqrt{2\pi} \times i \times S_{vmax} \tag{3}$$

$$V_s(\%) = \frac{A_s}{A_t} \times 100\% \tag{4}$$

where V_s is the volume loss and A_s is the area of induced settlement, i represent the point of inflection, S_{vmax} and A_t represent maximum ground settlement and the area of tunnel respectively.

Point of inflection (i) is the horizontal displacement to the point where the total displacement of that point area directed toward to a 'point sink' which assumed as the tunnel axis as shown on Fig. 9. In Fig. 9, the ground surface settlement vectors are extended towards the tunnel axis. The surface nodes that has the closest intersection to the tunnel axis is considered as the point of inflection (Potts 2017).

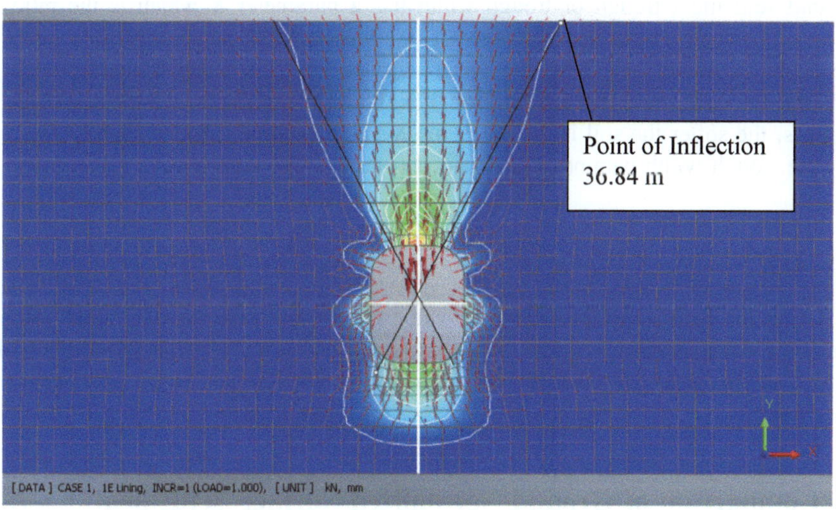

Fig. 9. Ground displacement vector on the final step

The volume loss in Yurakucho case is ranged between 0.1 and 0.2% by using Eqs. (3) and (4). Zhang (2011) has showed that the ground volume loss for Old Alluvium soil which has similar soil properties as Yurakucho site, is approximately around 0.1–0.9% as shown on Fig. 10. The calculated volume loss from the analysis results are plotted on the same figure.

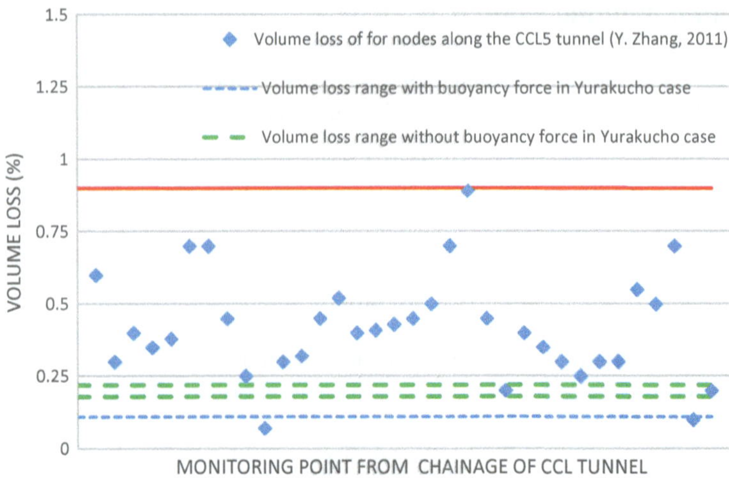

Fig. 10. Comparison of volume loss using EPB to old alluvium (Zhang 2011)

6 Ground Settlement Trough Ratio

Ground settlement trough or trough width has a parameter *K* which is the ratio of surface distance between the tunnel center-line position and the point of inflection *i* to the depth of tunnel axis from the tunnel surface z_0. This value is mainly used to calculate the Gaussian formulation. Trough width ratio *K* differs when the soil stiffness changes, the softer the soil is the lower the *K* value become (Moller 2006).

The trough width parameter is obtained from Eq. (5) as follows:

$$K = \frac{i}{z_0} \tag{5}$$

Using the case study field data, the value *K* with buoyancy is 0.543 and without buoyancy consideration is 0.591. This shows the buoyancy consideration affects the trough width ratio, as the *K* value reduces with the influence of buoyancy consideration.

7 Comparison of Ground Normalized Surface Settlement with Other Cases

In the following section, the normalized ground surface settlement of Yurakucho tunnel is compared with other tunnelling cases from previous literature. The other cases are DLR Tunnel in London and Heinenoord Tunnel in Netherland. The normalized surface settlement of these two tunnelling cases are shown in Fig. 11 (Tadashi and Soga 1999; Moller 2006).

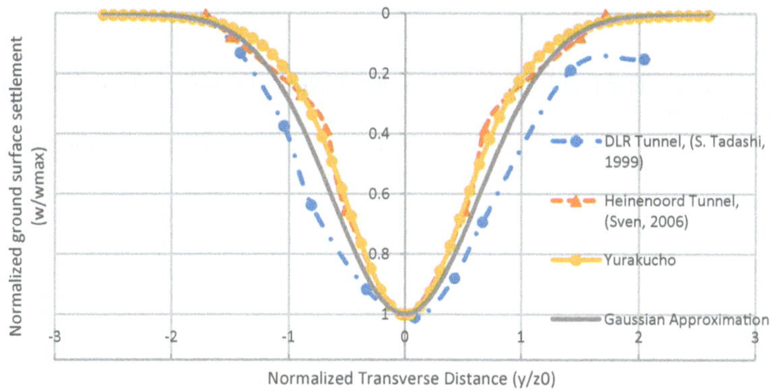

Fig. 11. The comparison of normalized ground settlement with other cases

The normalized ground settlement profile shows that the Yurakucho and Heinenoord tunnel cases have a close profile shape for w/w_{max} more than 0.5. This agreement is due to the similar close K value of the two cases. Yurakucho has $K = 0.591$, while the Heinennoord tunnel has $K \approx 0.64$. The reason why DLR Tunnel has a wider transverse distance as shown on Fig. 11 is due to the stiffer soil properties on the site. The DLR tunnel has $K \approx 0.77$ which indicate that it has a stiffer soil compared to the other case study.

8 Conclusions

In this paper, the lining analysis is done by studying the lining deformation and lining axial force and bending moment. The axial force and the bending moment have been compared with the field data. The analyses show how the shape and the magnitude of the bending moment and axial force satisfy the field measurement. The ground volume loss in Yurakucho tunnel case is calculated and compared with similar tunnel cases with similar condition. Yurakucho case is compared to other tunnelling cases from previous studies. Ground settlement trough were obtained and compared with Gaussian approximation. The normalized ground surface settlement of Yurakucho case is wider in comparison with the Gaussian approximation due to the soft ground in the Yurakucho site. The result shows how the analysis result have similar behaviour and the normalized graph shows how the Gaussian approximation behave with the change of K value. The curve with the higher K value created a wider trough curve. The change of the K value differs as the point of inflection i value changes. The higher point of inflection value is obtained as the soil is stiffer, as DLR tunnel has a stiffer soil on the site compared to Yurakucho and Heinennoord tunnel.

References

Moller, S.: Tunnel induced settlements and structure forces in linings. Doctoral thesis, University of Stuttgart (2006)

Tadashi, S., Soga, K.: Observation of ground movement during tunnel construction by slurry shield method at dockland light railway Lewisham extension. Jpn. Geotech. Soc. **39**(3), 99–112 (1999)

Verruijt, A., Strack, O.E.: Géotechnique **58**(6), 513–515 (2008). https://doi.org/10.1680/geot.2007.00046

Wan, M., Standing, J., Potts, D., Burland, J.: Measured short-term ground surface response to EPBM tunnelling in London Clay (2017). https://doi.org/10.1680/jgeot.16.p.099

Zhang, Y.H.: Volume Loss Caused by Tunneling of Circle Line Projects. Land transport Authority, Singapore (2011)

Sustainable Reconstruction in Conditions of Dense Urban Development

Talal Awwad[1,2(✉)], Vladimir Gruzin[3], and Vladimir Kim[1]

[1] L.N. Gumilyov Eurasian National University, Astana, Kazakhstan
dr.awwad.gfce@gmail.com
[2] Department of Geotechnical Engineering, Damascus University, Damascus, Syria
[3] S. Seifullin Kazakh Agro Technical University, Astana, Kazakhstan

Abstract. Sustainable Reconstruction in conditions of dense urban development is one of the urgent problems of modern construction. It is characterized by a significant increase in the load on existing foundations, the performance of all types of work in fairly difficult soil conditions and cramped situations that limit a performance of certain technological operations and an application of the necessary devices of mechanization. Therefore, a particular importance is referred to the using of advanced technologies that ensure minimum constructional, environmental and material risks, taking into account security measures, for both construction team and the residents in the area of construction. However, the current trend leads to a significant increase in the cost of reconstruction. So, innovative solutions that ensure a performance of required indicators of bearing capacity with reducing the cost of reconstruction works are undoubtedly relevant. This paper shows that the use of composite materials will significantly improve the reliability and durability of construction, as well as reduce the expenditure of building materials and the time of reconstruction. The paper also presents results of the research on development of rational technological and constructive solutions of reconstruction in dense building areas.

1 Introduction

Reconstruction of old buildings is one of the relevant problems associated with modern construction in the areas of congested urban development, it is characterized by Awwad et al. (2017), Долматов (1988), Абраменков and Грузин (1999), Awwad et al. (2016), Абраменков et al. (2012) and Грузин et al. (2014c, d, e) Patent № 28833, 28834, 29158:

(1) Considerable increase of loads applied to the existing foundations;
(2) Implementation of all types of works taking into account difficult soil conditions;
(3) Congested areas, which restrict implementation of some technological operations and application of the required mechanical tools:

   – There are different aboveground material objects and underground utilities in a construction site and in adjacent areas, which largely reduce the space of an operational zone;

M.-C. Weng et al. (eds.), *Current Geotechnical Engineering Aspects of Civil Infrastructures*, Sustainable Civil Infrastructures,
https://doi.org/10.1007/978-3-319-95750-0_2

- There is a lack of space for operation of excavating, building and road-making machines, and there is no access way for heavy vehicles during the workflow in a construction site;
- There are requirements to use progressive technologies, which provide minimal construction, ecological and material risks, taking into account safety measures for both construction workers and residents, who live in the area of implementation of construction works.

During the last years, many experimental, numerical, and analytical studies have been performed to investigate reinforced soil foundations—RSFs (Chen and Abu-Farsakh 2015); to investigate the behavior of reinforced soils and foundations for different reinforcement types: by use of soil cement columns and prefabricated vertical drains (Ishikura et al. 2016; Ye et al. 2015); stone columns (Killeen and McCabe 2014; Ng and Tan 2014); Lime pile (Abiodun and Nalbantoglu 2015; Awwad 2016); lime–cement columns (Larsson et al. 2009); by use of multiple-geocell or planar geotextile reinforcing layers (Tafreshi et al. 2016); by use of geogrid sheets (Chakraborty and Kumar 2014). In these terms, there are perspective technologies associated with improvement of soils adjacent to the existing urban buildings and structures (Абраменков et al. 2011; Awwad and Donia (2016); Zhussupbekov et al. 2013; Awwad and Al-Asali 2014).

Nowadays, in spite of the considerable annual pace of civil and residential construction in many cities and towns of the Republic of Kazakhstan, there are some unsolved problems associated with buildings and structures constructed in the 1950–70s, due to different reasons, they are to be demolished or, in the best-case scenario, renovated. This problem is very characteristic for transformation of the image of Astana, the capital of Kazakhstan, and its districts (the old city) constructed during the period of the USSR, when the city was called Tselinograd. The architectural appearance of the old city is mainly represented by 5-floor buildings with walls of 2–3 brick thickness or made of panels, which rest upon strip foundations. If 5-floor panel buildings, for example, those, located in Pobedy avenue, near Saken Seifullin Kazakh Agrotechnical University, are to be demolished; low-rise brick buildings can be successfully reconstructed with account of improvement of their soils and increase of bearing capacity of strip foundations.

2 The Special Features of Reconstruction of Buildings and Structures During Transformation of the Architectural Image of the City

Today, when solving problems associated with reconstruction of modern industrial, civil and residential buildings and structures, which become more difficult every year and they are characterized by considerable increase of loads applied to the foundations, a special role in the construction practice is given to application of progressive technologies of preparation of soils and construction of pile foundations. However, the existing trend leads to a significant increase of the cost of works in the field of foundation engineering, therefore, the innovative solutions, which provide achievement of the required values of bearing capacity and simultaneously reduce the prime cost of reinforcement of the earthworks of the existing urban development, are certainly very relevant.

At the same time foundation engineering has some reserves for increasing efficiency, decreasing estimated costs and improving quality due to:

- Implementation of progressive technologies of reinforcing different types of foundations of the existing buildings and structures with composite materials;
- Application of updated structures of pile foundations;
- Use of innovative technologies of hardening the existing subgrades of different industrial, civil and residential buildings and structures.

Nowadays a group of scholars is carrying out works for elaboration of rational technological and structural solutions for reconstruction of construction objects in the city of Astana, the Republic of Kazakhstan. Under the conditions of congested city development, the existing 2–3-floor residential buildings and separate structures rest on strip foundations. Therefore, in the framework of reconstruction of these structures it is suggested to realize a number of the following technological operations (Fig. 1) using composite materials in order to increase bearing capacity of the foundations as well as to enhance considerably reliability and durability of the earthworks. Moreover, it allows reducing consumption of construction materials and decreasing terms of reconstruction of the existing foundations of urban buildings and structures. In the view of the above mentioned the following technological sequence for increasing bearing capacity of strip foundations is proposed (Figs. 1 and 2).

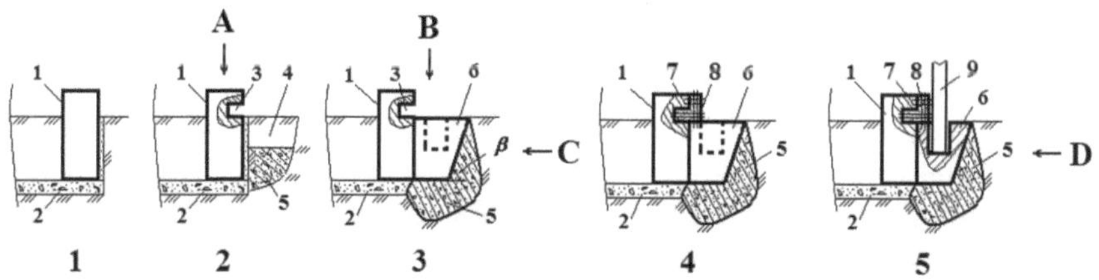

Fig. 1. The technological sequence of operations for reinforcement of the soil and strip foundation of a building in a congested urban area. Top view A and B (Side view C and D see Fig. 2). 1—strip foundation; 2—soil; 3—niche; 4—excavation in the subsoil; 5—rigid construction material; 6—RC block; 7—frame; 8—strut; 9—column

The method of reinforcement of the soil and foundations of buildings and structures under the conditions of congested urban development includes the following technological sequence of operations:

Position 1—the initial state of strip foundation 1 and soil 2 of a building or structure in a congested urban area is shown.

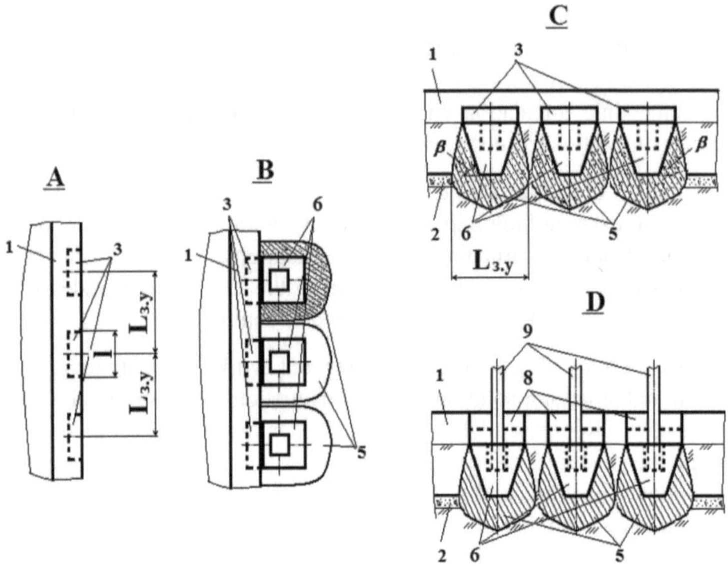

Fig. 2. Drawings A, B, C and D for positions 2, 3 and 5 of the technological sequence of operations for reinforcement of the soil and strip foundation of a building in a congested urban area. 1—strip foundation; 2—soil; 3—niche; 5—rigid construction material; 6—RC block; 8—strut; 9—column

Position 2—there are some preparatory operations before implementation of the technological sequence for reinforcement of soil 2 and foundation 1 of a building linked to (Fig. 2):

- making niches 3 of length l in the strip foundation equal to the width of the upper base of concrete block 6 at the distance $L_{3 \cdot y}$ between them;
- excavation 4 of soil under niche 3 near strip foundation 1, the volume of excavation is equal to the volume of concrete block 6;
- filling rigid construction material 5 (clay, crushed stone, concrete mixture etc.) into excavation 4.

Position 3—RC block 6 is installed in excavation 4 under niche 3 in strip foundation 1 and it is gradually tamped into rigid construction material 5 down to the designed elevation (Fig. 2) with account of forming the width of the consolidation area $L_{3 \cdot y}$ in soil 2 according to formula (Абраменков et al. 2012):

$$L_{3 \cdot y} = l \cdot \frac{K_{yn\pi} \cdot \rho_{\max}}{\left(K_{yn\pi} \cdot \rho_{\max} - \rho_0\right)}, \tag{1}$$

where ρ_0, $\rho_{\max}$—a density of soil before consolidation and its maximum value at optimum water content after consolidation, respectively;

$K_{yn\pi}$—a coefficient of soil consolidation while driving an RC block into it according to the condition:

$$\frac{\rho_1}{\rho_{\max}} = K_{yn\pi} \geq 0.95, \tag{2}$$

where ρ_1—a required density of the soil.

The RC block is made with the inclination angles of apothems of three lateral planes, which are equal to β or are more than the angle of internal friction ϕ (see Figs. 1 and 2):

$$\beta \geq \phi. \tag{3}$$

Position 4—frame 7 made of basalt rods or reinforcement is installed in niche 3 and in the space of RC block 6 for its attachment to strip foundation 1.

Position 5—cast formwork is made (it is not shown in Fig. 1) and strut element 8 is cast of concrete, after its curing column 9 is installed in the space of RC block 6 for further increasing the number of storeys in a building and providing the transfer of the required load to the foundation (see Figs. 1 and 2).

Implementation of the abovementioned sequence of technological operations for the technology of upgrading the building foundation under the conditions of congested urban development provides the required bearing capacity of the subsoil of the foundation due to:

(1) consolidation of the soil with rigid construction material, adjacent to the strip foundation, by jacking an RC block of the special geometry of cross and axial sections taking into account Conditions (1) and (2):
   - Figure 1 shows technological sequence 3;
   - Figure 2—top and side views—B, C;

   there are demonstrated the newly formed consolidated zones made of construction material 5, they were formed due to jacking an RC block into a strut,

(2) rigid connection of the existing strip foundation with a newly formed one using special strut element 8 (see Fig. 1—technological sequences 4 and 5).

The technology of consolidation of soil with rigid construction material by means of jacking an RC block ensures creation of a consolidated area under the whole combined foundation that provides general reduction of consumption of construction materials during reconstruction of buildings and structures in the congested urban areas.

The technical result of the following technological sequence is reached due to decreasing costs of reinforcement of the foundation of a building or structure while jacking RC blocks of trapezoid cross-section along its perimeter and forming consolidated soil areas around them. Later they are attached to the existing foundation with the strut elements, which are made using composite materials, and the columns are installed in them in order to increase the number of storeys of a building.

It is noteworthy, the relevance of applying composite and other reinforcing materials for underpinning foundations and other construction concrete structures is doubtless, as composites based on continuous basalt fiber have the following features, compared, for example, to steel reinforcement of grade A-Ш (A400C) (A-Sh (A400S)) (Ларин et al. 2012; Демешкин and Шваб 2011):

- an ultimate tensile strength is 3.5–4 times bigger;
- a thermal conductivity is 100–150 times smaller;
- there is a corrosive resistance to aggressive medium, including the alkaline medium;
- environmental friendliness is provided due to the absence of harmful and toxic substances;
- a predicted durability is minimum 100 years.

Numerous scientific studies have found that the use of continuous basalt fiber as the main component while making various concrete structures provides not only the significant increase of their strength but also the reduction of their weight, therefore, decreasing consumption of construction materials by 18–32%.

Due to the fact that in some cases there is no possibility to use standard RC blocks as reinforcing elements of the existing foundations because of some structural features of an existing building or a proposal of its reconstruction, it is suggested to use the technology of construction cast-in-place piles in the boreholes, which have been formed while jacking of working tools of special geometry or drilling holes with a special working tool (Абраменков et al. 2006; Грузин 1999; Грузин et al. 2014a, b Patent № 28832, 28835).

3 Definition of Coefficient of Correlation Between Elements of Working Tools

The rational correlation of the parameters of basic elements of a jacked working tool, including a height of a tip H_H, a body H_K, a bell end, a head $H_{p,o}$, and the angles (α, β, γ) of tilt of planes apothems to the axis of these elements allows achieving a maximum unit bearing capacity of pile foundations on the consolidated soil in different soil conditions. Depending on physical-mechanical properties of a soil mass of a construction site, structural features of the existing foundation and its final reconstruction for obtaining the required soil density while forming a borehole, a structure of the working tool can include a bell end, a head or nothing (Грузин et al. 2003).

The coefficient of correlation between the heights of the tip k_H, the body k_K and the bell end (head) $k_{p,o}$ of the working tool, which provide the required bearing capacity of a cast-in-place pile, are presented by the following dependencies:

$$k_H = \frac{H_H}{H}, \quad k_K = \frac{H_K}{H}, \quad k_{p,o} = \frac{H_{p,o}}{H}, \qquad (4)$$

where H—a height of the working tool;

$H_H, H_K, H_{p,o}$—heights of the tip, the body and the bell end (head), respectively.

As the maximum soil density in the borehole wall depends on a value of its deformation, which, in its turn, is defined by a value of stresses applied to the contact surface of the plate and soil mass, according to Condition (2), the maximum soil density is provided in the border of soil contact with the lateral surface of the working body. Consequently, according to the condition of soil weight preservation, we get the ultimate unit bearing capacity of a future cast-in-place pile.

In order to obtain the ultimate unit bearing capacity of a future cast-in-place pile and the maximum radial area of consolidation of the soil mass around it, based on the condition of soil weight preservation, we define its maximum density in the border of soil contact with the lateral surface of the trapezoid tamping working tool. Based on the given assumption we define (Абраменков et al. 2006):

$$D_H = D_{3\cdot y} \cdot \sqrt{\frac{K_{yn\pi} \cdot \rho_{\max} - \rho_0}{K_{yn\pi} \cdot \rho_{\max}}}, \tag{5}$$

where ρ_0, $\rho_{\max}$—a soil density before its consolidation and the maximum soil density, respectively;

$D_{3\cdot y}$—a diameter of the inscribed circle of a consolidated area of the soil mass while driving a trapezoid working tool into it;

D_H—a diameter of the inscribed circle in the lower base of the body of a working tool;

$K_{yn\pi}$—a coefficient of soil consolidation.

Formula (5) establishes the correlation between a diameter of the inscribed circle in the lower base of the body of a working tool, a diameter of the inscribed circle of a consolidated area of the soil mass while driving a trapezoid working tool into it and soil mass properties. Taking into account congruence (5) there have been defined coefficients of correlations between heights of structural elements of the working tool (k_K, k_H and $k_{p,o}$) and a height of the working tool H:

- for the tip:

$$k_H = \frac{D_{3y}}{2 \cdot H \cdot \mathrm{tg}\left(\frac{\alpha}{2}\right)} \cdot \sqrt{\frac{K_{yn\pi} \cdot \rho_{\max} - \rho_o}{K_{yn\pi} \cdot \rho_{\max}}}, \tag{6}$$

where k_H—a coefficient of correlation between a tip height and a height of the working tool

α—a cutting-point angle of the trapezoid working tool;

- for the body:

$$k_K = \frac{D_K - D_{3y}\sqrt{\frac{K_{yn\pi} \cdot \rho_{\max} - \rho_o}{K_{yn\pi} \cdot \rho_{\max}}}}{2 \cdot H \cdot tg\beta} \tag{7}$$

where k_K—a coefficient of correlation between a body height and a height of the working tool;

β—an angle of inclination of the plane apothem of the trapezoid body of the working tool to its axis;

- for the bell end or the head:

$$k_{p,o} = \frac{K_o \cdot D_H}{2 \cdot H \cdot tg\beta} \tag{8}$$

where $k_{p,o}$— a coefficient of correlation between a bell end (head) height and a height of the working tool;

K_o—a coefficient of relative soil deformation.

Based on the conducted analytical research these have been developed engineering methods of selection and calculation of structural parameters of elements of the working tool, which have been implemented during design, production and exploitation of trapezoid working tools of the equipment for making boreholes for cast-in-place piles which rest upon the consolidated soil:

- with the bell end (Fig. 3a);
- with the head (Fig. 3b);
- without the bell end and the head (Fig. 3c).

(a) (b) (c)

Fig. 3. The structures of the working bodies for making boreholes in soil: **a** with the bell end; **b** with the head; **c** without the bell end and the head

The abovementioned working tools have been applied for construction of civil, industrial and residential buildings and structures.

Taking into account technical requirements for a project of works on a foundation construction and depending on physical and mechanical properties of soil in a construction site, there is used one of the following types of the working tools for soil compaction in the bottom, walls or head of a formed borehole (see Fig. 4).

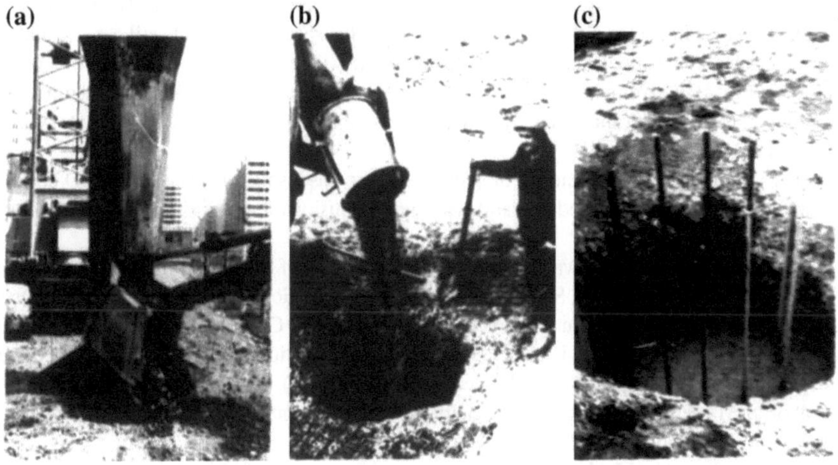

Fig. 4. Construction of a cast in place pile in soil, which rests on the consolidated subsoil: **a** forming a consolidated borehole by the working tool along with filling rigid construction material; **b** filling the newly formed borehole with commercial concrete; **c** installation of the a reinforced frame

4 Conclusions

A successful solution of the problems of reconstruction of residential and civil buildings and structures in congested urban areas is conditioned by a whole range of factors, which should be taken into account before implementation of such types of works. To consider their capabilities, it is necessary to conduct a comparative analysis to substantiate the applied progressive technologies of reinforcement of the existing strip foundations with account of using the contemporary mechanical tools, which provide preservation of structures adjacent to a construction site. Another important aspect is to define ability of soils adjacent to the foundations under reconstruction to harden by means of consolidation that will allow increasing bearing capacity of the foundation under reconstruction according the plan of workflow.

References

Abiodun, A.A., Nalbantoglu, Z.: Lime pile techniques for the improvement of clay soils. Can. Geotech. J. **52**(6), 760–768 (2015)

Awwad, T.: Problems of Syrian historical monuments, destroyed by military action: diagnostics, strengthening and reconstruction, challenges and innovations in geotechnics. In: Proceedings of the Eighth Asian Young Geotechnical Engineering Conference, pp. 21–27 (2016)

Awwad, T. Al-Asali, B.E.: Efficiency of improving the specifications of soil lenses that are formed near the tunnel during the stages of its investment. In: Proceedings of the TC207 ISSMGE Conference, Soil-Structure Interaction, Underground Structures and Retaining Walls. Advances in Soil Mechanics and Geotechnical Engineering, vol. 4, pp. 111–118 (2014)

Awwad, T., Donia, M.: The efficiency of using a seismic base isolation system for a 2D concrete frame founded upon improved soft soil with rigid inclusions. Earthq. Eng. Eng. **15**(1), 49–60 (2016)

Awwad, L., Rahme, N., Awwad, T.: The rehabilitation of historic buildings—case-study Al-Thakana Al-Hamidieh in Syria, challenges and innovations in geotechnics. In: Proceedings of the Eighth Asian Young Geotechnical Engineering Conference, pp. 297–302 (2016)

Awwad, T., Donia, M., Awwad, L.: Effect of a stiff thin foundation soil layer's depth on dynamic response of an embankment dam. Procedia Eng. **189**(2017), 525–532 (2017)

Chakraborty, M., Kumar, J.: Bearing capacity of circular foundations reinforced with geogrid sheets. Soils Found. **54**(4), 820–832 (2014)

Chen, Q., Abu-Farsakh, M.: Ultimate bearing capacity analysis of strip footings on reinforced soil foundation. Soils Found. **55**(1), 74–85 (2015)

Ishikura, R., Yasufuku, N., Brown, M.: An estimation method for predicting final consolidation settlement of ground improved by floating soil cement columns. Soils Found. **56**(2), 213–227 (2016)

Killeen, M.M., McCabe, B.A.: Settlement performance of pad footings on soft clay supported by stone columns: a numerical study. Soils Found. **54**(4), 760–776 (2014)

Larsson, S., Rothhämel, M., Jacks, G.: A laboratory study on strength loss in kaolin surrounding lime–cement columns. Appl. Clay Sci. **44**(1–2), 116–126 (2009)

Ng, K.S., Tan, S.A.: Design and analyses of floating stone columns. Soils Found. **54**(4), 478–487 (2014)

Tafreshi, S.N.M., Sharifi, P., Dawson, A.R.: Performance of circular footings on sand by use of multiple-geocell or planar geotextile reinforcing layers. Soils Found. **56**(6), 984–997 (2016)

Ye, G., Zhang, Q., Zhang, Z., Chang, H.: Centrifugal modeling of a composite foundation combined with soil-cement columns and prefabricated vertical drains. Soils Found. **55**(5), 1259–1269 (2015)

Zhussupbekov, A., Zhunisov, T., Issina, A., Awwad, T.: Geotechnical and structural investigations of historical monuments of Kazakhstan. In: Geotechnical Engineering for the Preservation of Monuments and Historic Sites, Proceedings of 2nd International Symposium on Geotechnical Engineering for the Preservation of Monuments and Historic Sites, pp. 779–784 (2013)

Абраменков, Д.Э., Абраменков, Э.А., Грузин, А.В., Грузин, В.В.: Строительные машины и оборудование в фундаментостроении. Учебник. Астана, Фолиант, 296 с (2011)

Абраменков, Д.Э., Грузин, А.В., Грузин, В.В.: Средства механизации и технология строительного производства. Монография. Под общ. ред. д.т.н., проф. Э.А. Абраменкова. – Saarbrucken. Palmarium Academic Publishing, Germany, 327 с (2012)

Абраменков Д.Э., Абраменков Э.А., Грузин В.В. Средства механизации строительства, реконструкции и реставрации зданий, сооружений. Учебник. Новосибирск, НГАСУ (Сибстрин), 320 с (2006)

Абраменков, Э.А., Грузин, В.В.: Средства механизации для подготовки оснований и устройства фундаментов. – Новосибирск, Новосибирский государственный архитектурно-строительный университет, 215 с (1999)

Грузин, В.В.: Типоразмерные ряды средств механизации для подготовки оснований и устройства свайных фундаментов. /Изв. Вузов. Строительство. №9. С. 59–64 (1999)

Грузин, В.В., Бушманова, Е.В., Абраменков, Д.Э.: Обоснование и выбор рациональных параметров рабочего органа для вытрамбовывания котлованов. /Известия ВУЗов. Строительство №6. С.51–55 (2003)

Грузин, В.В., Нураков, С.Н., Жантлесов, Ж.Х., Грузин, А.В.: Инновационный патент № № 28832. Гидромолот импульсный вдавливающий. Астана, Комитет по правам интеллектуальной собственности МЮ РК, опубл. 15.08.2014, бюл. № 8 (2014a)

Грузин, В.В., Нураков, С.Н., Жантлесов, Ж.Х., Грузин, А.В.: Инновационный патент № № 28835. Навесное оборудование для раскатывания скважин сложного профиля. Астана, Комитет по правам интеллектуальной собственности МЮ РК, опубл. 15.08.2014, бюл. № 8 (2014b)

Грузин, В.В., Нураков, С.Н., Жантлесов, Ж.Х., Грузин, А.В.: Инновационный патент № 29158. Способ усиления грунтовых оснований фундаментов зданий и сооружений, от 17.11.2014, бюл. №11 (2014c)

Грузин, В.В., Нураков, С.Н., Жантлесов, Ж.Х., Грузин, А.В.: Инновационный патент № 28834. Способ усиления грунтовых оснований и фундаментов зданий и сооружений, от 15.08.2014, бюл. № 8 (2014d)

Грузин, В.В., Нураков, С.Н., Жантлесов, Ж.Х., Грузин, А.В.: Инновационный патент № 28833. Способ усиления фундаментов зданий и сооружений, от 15.08.2014, бюл. № 8 (2014e)

Демешкин, А.А., Шваб, А.А.: Экспериментальное исследование механических свойств непрерывного базальтового волокна применительно к производству композиционных материалов. Вестник Самарского гос.техн. университета, Сер. Физ-мат наук №3 (24) С.185–188 (2011)

Долматов, Б.И.: Механика грунтов, основания и фундаменты. Л., Стройиздат - 397 с (1988)

Ларин, А.А., Мерзляков, Г.А., Пономарев, В.Б., Рафаилов, А.Г., Резниченко, В.И.: Особенности применения изделий из базальтового непрерывного волокна в строительстве, журнал «Прогрессивные технологии в системе машиностроения», вып.1, 2 (44) (2012)

Influence of Unit Weight of Slurry-Deposited Coal Ash on Stability of Ash Dykes Raised by Upstream Method

Aali Pant[1(✉)], Manoj Datta[2], and Gunturi Venkata Ramana[2]

[1] IIT Delhi, Delhi, New Delhi, India
aali.pant@gmail.com
[2] Civil Engineering Department, IIT Delhi, Delhi, New Delhi, India

Abstract. In this paper efforts have been made to understand the influence of unit weight of slurry-deposited coal ash on the slope stability of ash dykes raised by upstream method. Using the geotechnical properties of compacted ash as well as slurry-deposited ash, a series of stability analysis have been carried out to identify the critical factor of safety for incrementally raised ash dykes. A 19 m high ash pond in Eastern India had been incrementally raised to its current elevation, by upstream method in four raisings. The ash pond was to be further raised by providing two raisings (5th and 6th) of 3 m each making the ash pond 25 m high.

Slope stability analysis of downstream slope of the dykes was carried out for steady state seepage condition using effective stress parameters for the ash as well as foundation soil. For the 5th and 6th raising the critical failure surface was observed to pass predominantly through slurry deposited ash and the foundation soil beneath it. The factor of safety (FOS) was observed to be low on account of the low unit weight as well as the low shear strength of ash. It was observed that for saturated unit weight (γ_{sat}) of 11.7 kN/m^3 the FOS was lower by 12.5% in comparison to γ_{sat} of 16 kN/m^3 for 5th raising and by 14% for 6th raising in static condition. Similarly, in seismic condition, the FOS at γ_{sat} of 11.7 kN/m^3 was observed to be low by 14% than that at γ_{sat} of 16 kN/m^3 for 5th raising and by 15% for 6th raising. The role of unit weight in influencing slope stability of incrementally raised dykes of slurry ponds has been reported for the first time in this study and it shows that light-weight deposits such as coal ash can lead to low FOS when the failure surface passes through the slurry-deposited material.

1 Introduction

According to the data from World Bank on an average 31.5% of electricity production is from coal source. Figure 1 shows the distribution of percentage of electricity generated from coal in the different region of the world till the year 2014. This leads to massive production of coal ash annually. The utilization of coal ash is being encouraged throughout the world, but a large percentage of coal ash still remains unutilized. This unutilized ash is stored in ash ponds through either wet disposal or in mounds through dry disposal. Coal ash storage ponds are complex geotechnical structures. Coal ash is discharged, in the form of slurry, from pipes located in the perimeter of the ash

M.-C. Weng et al. (eds.), *Current Geotechnical Engineering Aspects of Civil Infrastructures*, Sustainable Civil Infrastructures,
https://doi.org/10.1007/978-3-319-95750-0_3

pond. Depending on the slurry deposition method the slurry usually has solid to liquid ratio ranging from 1.5:1 in volume for high concentration slurry deposition to 1:6 for low concentration slurry deposition. Part of the water infiltrates into the deposited ash, while the rest flows as surface runoff down to the decanting wells.

Fig. 1. Percentage of electricity generated from coal in different regions of the world. *Source* IEA Statistics © OECD/IEA 2014 (iea.org/stats/index.asp)

There are numerous failures of coal ash ponds reported in past. The failure of the Tennessee Valley Authority's Kingston Fossil plant, in 2008 led to deterioration of community health. There are several other examples of slurry pond dam failures including tailings dams around the world. Hence, the design and construction of geotechnical earth structures like ash ponds requires assessment of their response to different properties of the slurry as well as response of the structure under static and dynamic loading conditions.

Apart from coal ash there are other industrial waste products which are disposed through slurry deposit in artificially made ponds. Mine tailings are similar by-products of industries that are hydraulically deposited in dykes. The stability of such dykes under static loading mainly depends upon the shear strength of the compacted dykes and the hydraulically deposited fill material. In the present study, the influence of unit weight of hydraulically deposited coal ash has been studied on the overall stability of an ash dyke raised by upstream method in four raisings. The stability analysis has been performed using the procedures outlined in the subsequent sections with Slope/w software packages for steady-state seepage condition.

2 Literature Review

A significant body of research can be found in the literature for laboratory methods to determine minimum and maximum dry unit weight of coal ash. The laboratory results on the minimum dry unit weight studied from literature are presented in Table 1. From literature review, it was observed that the range of minimum and maximum dry unit weight of coal ash deposited in ash ponds varies from 7.7–10.5 to 11.9–13.3 kN/m^3, respectively (Gray and Lin 1972; Toth et al. 1988; Dayal 1989; Gandhi et al. 1999; Horiuchi et al. 2000; Pandian 2004; Kim et al. 2005; Bachus and Santamarina 2014). The corresponding dry unit weight at 30% relative density varies from 8.6 to 11.2 kN/m^3. The minimum saturated unit weight has been calculated from the specific gravity values reported in the literature or by using representative values if the same have not been reported. The corresponding minimum saturated unit weight varies from 11.7 to 15.8 kN/m^3 over the entire range of dry unit weight. The low unit weight is attributed to low specific gravity of coal ash which varies from 1.90 to 2.59 (Pandian 2004; Das and Yudhbir 2005; Mishra and Karanam 2006; Jakka et al. 2010).

Table 1. Minimum dry unit weight of coal ash from literature review

Material	Minimum dry unit weight, kN/m^3	Minimum saturated unit weight, kN/m^3	References
Raichur Pond Ash	9.7	14.5	Pandian (2004)
Rae Bareili Pond Ash	9.4	14.1	
Neyveli Pond Ash	7.7	11.7	
Vijaywada Pond Ash	8.3–9.6	13.2–15.0	
Hekinan Fly Ash	8.73	13.1	Porbaha et al. (2000)
Matsushima Fly Ash	8.48	12.7	
Badarpur Pond Ash	10.5	15.8	Jakka et al. (2010)
Indraprastha Pond Ash	9.8	15.0	
Mettur Pond Ash	9.27	14.4	Gandhi et al. (1999)
Pond Ash	8.0–8.4	14.0–14.7	Pant et al. (2016)
Pond Ash	7.8–8.0	11.8–12.0	Sridharan (2012)
Bottom Ash	8.89	13.5	Seals et al. (1972)

Limited studies are reported in literature on in situ density of pond ash (Gandhi et al. 1999). Measurement of in situ field density of coal ash ponds is difficult due to difficulty in retrieving of undisturbed samples from boreholes. The N-values obtained from Standard Penetration Tests (SPT) are used to describe the density condition of soil. The N-values at ash ponds vary over a wide range from a minimum of 2 at shallow depths of 1.5 m to 11–15 at deep depths of 9–19 m (Jakka et al. 2011). A number of correlations are available to determine unit weight from N-values. However, many of these correlations are questionable as they are based on a small database or specific soils.

3 Background

A 19 m high ash pond in Eastern India had been incrementally raised to its current elevation, by upstream method in four raisings as shown in Fig. 2. The ash pond is to be further raised by providing two raisings (5th and 6th) of 3 m each making the ash pond 25 m high from ground level. 4 boreholes (BH-1 to BH-4) were drilled in the ash pond to a depth of 24 m so as to penetrate at least 5 m into the underlying local soil. In addition to that 6 boreholes (BH-5 to BH-10) were drilled around the periphery of the ash pond beyond the toe drain of the starter dyke to a depth of 15 m. Figure 3 shows the location of the 10 boreholes drilled in the ash pond and around its periphery. N-values as well as disturbed representative samples of ash and foundation soil were thus obtained. The ash samples were classified as sandy silt with N-values ranging from 5 to 18. Foundation soil was observed to be clayey silt/silty clay with N-values ranging from 5 to 30. From the boreholes drilled along the periphery of the ash pond, the ground water table was observed to be shallow being 1 m below the natural ground level.

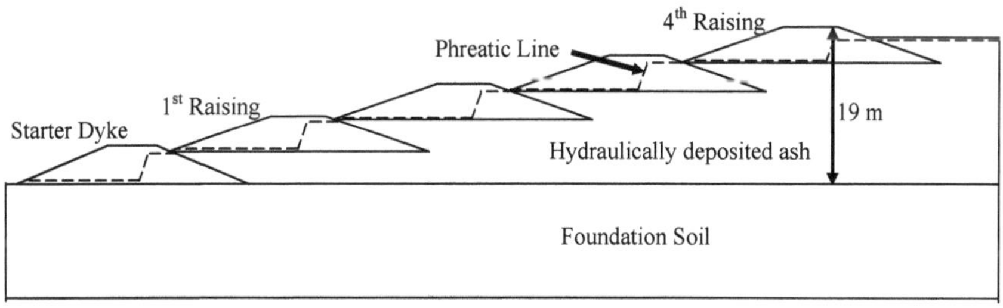

Fig. 2. Sectional view of the existing ash dyke

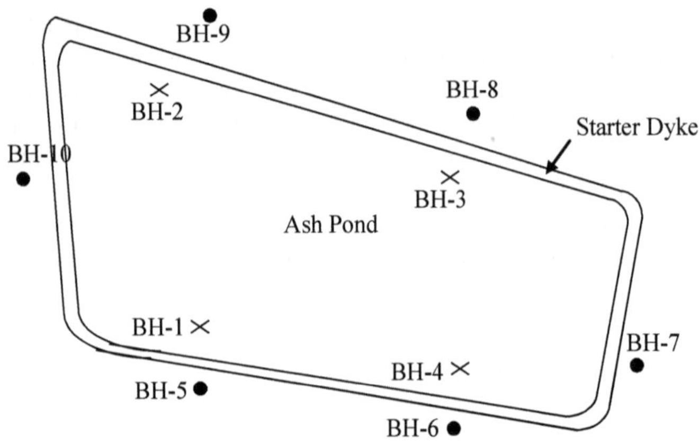

Fig. 3. Plan of the ash pond and location of the boreholes drilled

4 Experimental Investigation

Ash samples were collected from a depth of 0.5 m below the existing ash level from the ash pond. Representative ash samples were tested for complete geotechnical characterization. The ash was non-plastic in nature. Specific gravity value of ash was obtained to be 2.08 which is considerably lower than that of the natural soils. Grain size distribution of the ash samples was studied by both sieve analysis and hydrometer analysis. The ash samples had predominantly silt content (with 30% sand) and were classified as sandy silt. Compaction characteristics of the ash samples were studied using standard Proctor as per the Indian standard specifications. Effective-stress parameters, cohesion (c′) and angle of shearing resistance (ϕ′), of the ash samples were evaluated by performing direct shear tests on saturated samples in the loose and dense state. Loose samples were prepared by slow pouring and then saturating while dense samples were prepared by rodding and tamping and then saturating.

In addition to ash samples, local soil samples were also collected from a depth of 1.0 m below the existing ground level from two locations. Almost 100% of the local soil particles were finer than 75 μ sieve size. From the Atterberg limit tests conducted on the local soil it was observed that the liquid limit, w_l, of the soil was 65% and plastic limit, w_p, was 25%. The plasticity index, I_p, was thus 40%. The local soil was classified as clay of high plasticity (CH). The specific gravity of the soil was 2.49. The results of experimental investigation on the ash samples as well as local soil samples are reported in Table 2.

Table 2. Engineering properties of ash and soil samples

Property	Ash sample	Local soil
Sand size (4.75–0.075 mm), %	28	0
Silt size (0.075–0.002 mm), %	72	61
Clay size (<0.002 mm), %	0	39
Specific gravity	2.08	2.49
Maximum dry unit weight, kN/m^3	11.9	15.4
OMC, %	28.9	24
Minimum dry unit weight, kN/m^3	9.52	–
Liquid Limit, w_l, %	–	65
Plastic Limit, w_p, %	Non plastic	25
Plasticity Index, I_P, %	–	40
Angle of shearing resistance, ϕ, °	Loose (slurry deposited): 29	22
	Dense (compacted): 34	

5 Selection of Ash & Soil Parameters for Analysis

Figure 4 gives the SPT N-value for BH-1 to BH-4 drilled in the ash pond. BH-3 shows unusually low N-values. However, an average N-value of 7 was obtained from the field SPT tests conducted on the ash pond. Using the empirical method for computing ϕ', relative density and unit weight of granular soils based on the SPT N-value it can be inferred that medium dense sand with N-value of 7 has ϕ' of 30° (Meyerhof 1959).

From laboratory shear strength tests carried out on saturated ash samples, an angle of shearing resistance of 29° was obtained in loose state (30% relative density). Thus ϕ' value of 29° was used for the hydraulically deposited soil in stability analysis. Direct shear tests conducted on saturated sample of dense ash gave ϕ' value of 34°. The same was used as the shear strength parameter for the compacted ash dykes. Figures 5 and 6 show the variation of shear stress and vertical displacement with horizontal displacement for the pond ash samples at 30% relative density (Fig. 5a, b) and 70% relative density (Fig. 6a, b). The behaviour of the ash samples typically resembled those of a sandy soil. Granular material when sheared show slight initial volumetric contraction followed by volumetric expansion i.e. dilation (Bolton 1986). The ash samples exhibited strain-softening behaviour at dense condition with relative density of 70% while strain-hardening behaviour at loose condition with relative density of 30%.

The angle of shearing resistance of foundation soil has been estimated from correlations given by several researchers (Bjerrum and Simons 1960; Kanja and Wolle 1977; Ladd et al. 1977; Olsen et al. 1986). Corresponding to I_p of 40%, the effective angle of shearing resistance of the local soil has been estimated as 22°.

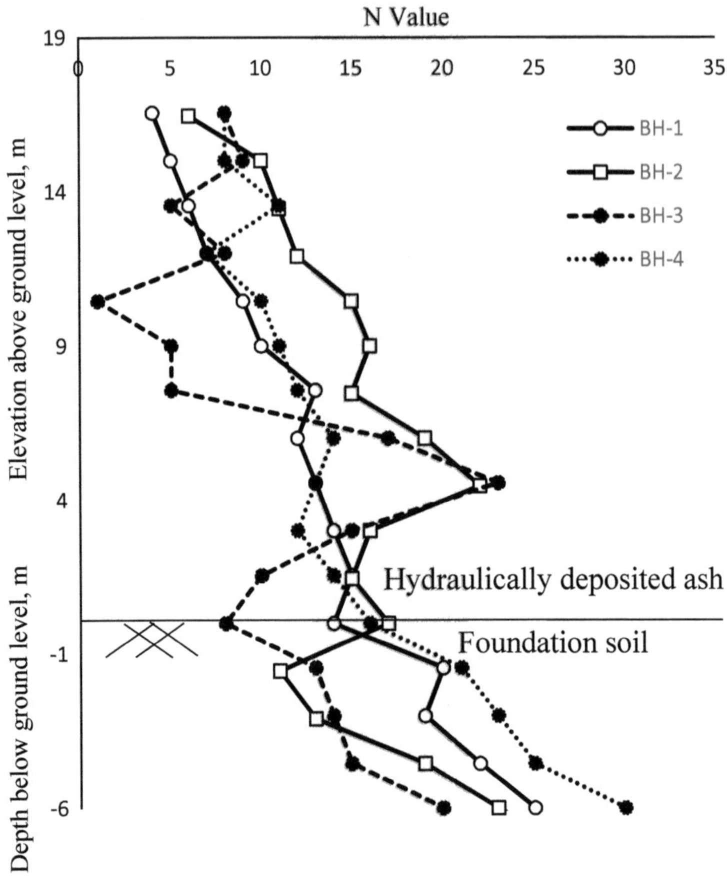

Fig. 4. SPT profile of boreholes drilled in the ash pond

6 Slope Stability Analysis

Ash pond dykes are raised by upstream method in order to enhance the volume of the pond. Since the pond ash is abundantly available on the construction site itself, it is economical to use ash as the construction material for raising the ash dyke embankment. Top cover of natural earth is provided to take care of erosion and gully formation. In the present stability analysis, a four-stage dyke incrementally raised by upstream method on the starter dyke was considered as base case. The analysis has been carried out for steady state seepage condition. A phreatic line (piezometric line) has been assigned on the basis of highest slurry level in the ash pond and the configuration of chimney drain and blanket drain of each raising; the software evaluates the pore water pressure on the basis of the location of the phreatic line. Following are the salient features of the ash dykes:

Starter dyke was constructed by local clayey soil with a downstream slope of 2.5H:1 V.

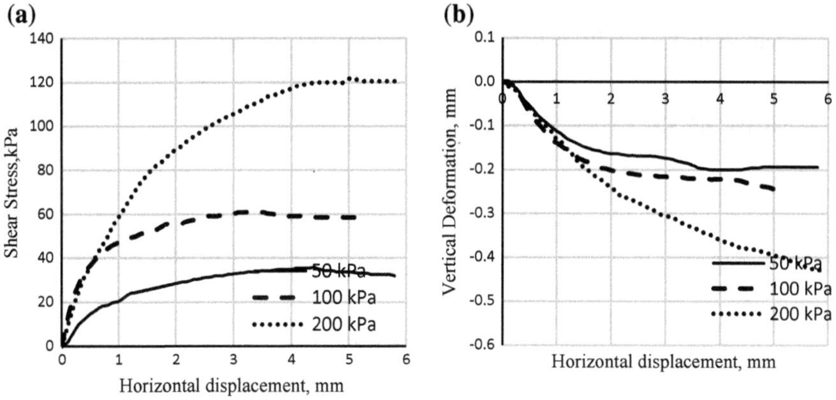

Fig. 5. Variation of shear stress and vertical displacement with horizontal displacement for the ash samples at 30% relative density

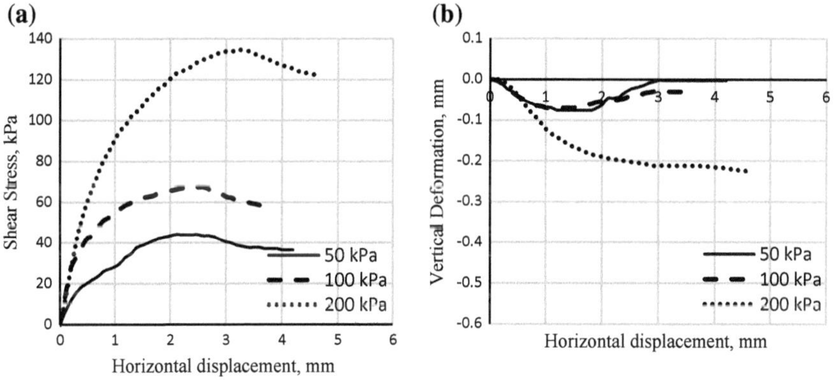

Fig. 6. Variation of shear stress and vertical displacement with horizontal displacement for the ash samples at 70% relative density (**a** & **b** respectively)

The downstream slope of all the raising have equal inclination on 3H:1 V.

The starter dyke is 6 m high while the further four raisings above starter dyke are of variable height. At its current elevation with four raisings, the total height of the dyke is 19 m.

The crest width of each raising is of 6 m.

The proposed 5th and 6th raising are 3 m high each, thus making the ash pond 25 m high from the existing ground level.

The issue of stability of an ash dyke with reference to the variability of unit weight of ash deposited in pond was the focus of the study. Slope stability analysis of downstream slope of the dykes was carried out for steady state seepage condition using effective stress parameters for the ash as well as foundation soil. The stability was checked for static as well as seismic condition. Keeping the angle of shearing resistance of ash constant, the influence of unit weight of ash was identified by varying saturated unit weight of hydraulically deposited ash from 11.7 to 16 kN/m^3 in steps of 1 kN/m^3.

7 Results of Stability Analysis

Slope stability analysis was performed for both static and seismic conditions using GEOSLOPE version Slope/W 2007. As per IS 1893, the ash pond lies in earthquake Zone III. The design values of horizontal seismic coefficient and vertical seismic coefficient have been calculated using seismic coefficient method and taken as 0.106 and 0.053 respectively. The minimum acceptable factor of safety (FOS) for steady state seepage condition in static and with seismic loading condition are kept 1.5 and 1.0 respectively as per IS-7894 (1975). The factor of safety (FOS) obtained for the 5th and 6th raising are listed in Table 3.

Table 3. FOS obtained from stability analysis

Saturated unit weight of slurry deposited ash γ_{sat}, kN/m^3	Static case		Seismic case	
	5th raising	6th raising	5th raising	6th raising
11.7	1.44	1.42	0.88	0.86
12.5	1.56	1.50	0.92	0.89
14	1.64	1.60	0.98	0.95
15	1.67	1.63	1.01	1.00
16	1.67	1.65	1.03	1.01

7.1 Static Condition

The factor of safety (FOS) obtained for the 5th and 6th raising in static condition are listed in Table 2. With the 5th raising, when the unit weight of slurry deposited ash was 11.7 kN/m^3 it was observed that the ash dyke was susceptible to failure with FOS being less than 1.5. With an increase in saturated unit weight of deposited ash, the FOS increases and the dyke became stable (FOS > 1.5). For unit weight of 11.7 kN/m^3 the FOS was lower by 12.5% in comparison to that at unit weight of 16 kN/m^3 for 5th raising.

With the 6th raising the FOS obtained was lower by 14% with a unit weight of 11.7 kN/m^3 than that at 16 kN/m^3. It was observed that the ash dyke was susceptible to failure at saturated unit weight of 11.7 and 12.5 kN/m^3. With an increase of saturated unit weight, the ash dyke became stable. The failure surface passes primarily through the slurry deposited ash and the foundation soil beneath it. Typical failure surfaces for static condition are shown in Figs. 7 and 8.

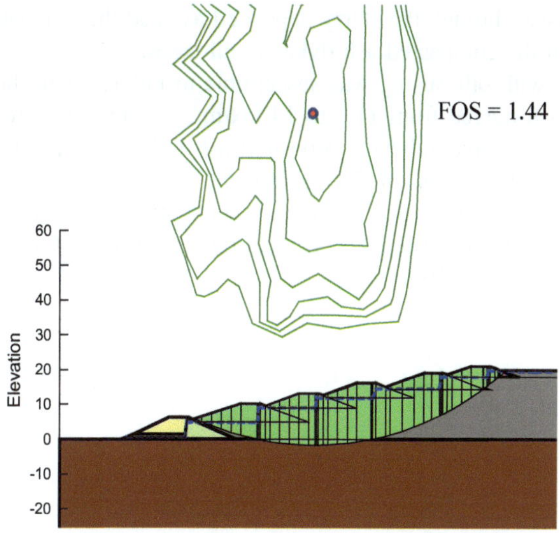

Fig. 7. Critical failure surface after 5th raising in static condition (γ_{sat} 11.7 kN/m^3)

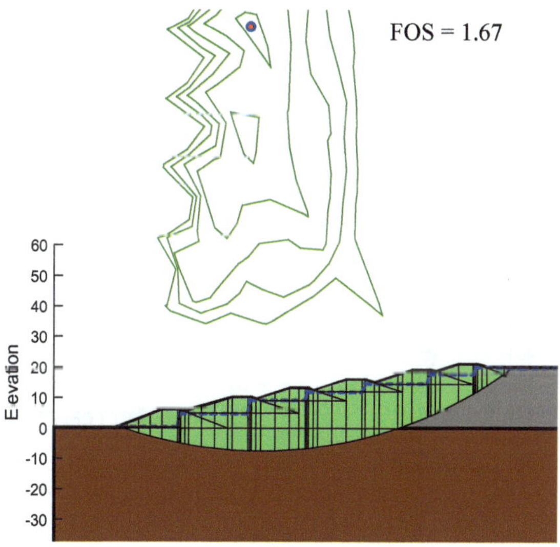

Fig. 8. Critical failure surface after 5th raising in static condition (γ_{sat} 15 kN/m^3)

7.2 Seismic Condition

The factor of safety (FOS) obtained for the 5th and 6th raising in seismic condition (pseudostatic method) are listed in Table 3. In seismic condition, the FOS at unit weight of 11.7 kN/m^3 was observed to be low by 14% than that at 16 kN/m^3 for 5th raising. The ash dyke was susceptible to failure when the unit weight of slurry deposited ash varied from 11.7 to 14 kN/m^3 with FOS of less than 1.0. The critical failure surface passes through the slurry deposited ash and the foundation soil beneath it and not through the compacted ash dykes in all cases.

The ash dyke with 6th raising was susceptible to failure when the unit weight of slurry deposited ash ranged from 11.7 to 14 kN/m^3. The FOS lowered by 15% when the unit weight of the hydraulically deposited ash reduced from 16 to 11.7 kN/m^3. Similar to the stability results for 5th raising, the critical failure surface passed primarily through the slurry deposited ash and the foundation soil beneath it, independent of the unit weight of the deposited ash. Typical failure surfaces of the stability analysis can be seen in Figs. 9 and 10.

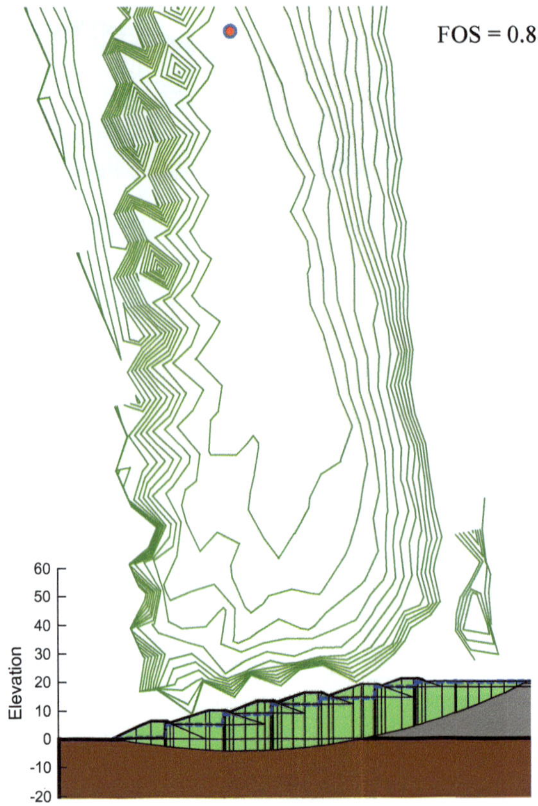

Fig. 9. Critical failure surface after 5th raising in seismic condition (γ_{sat} 11.7 kN/m^3)

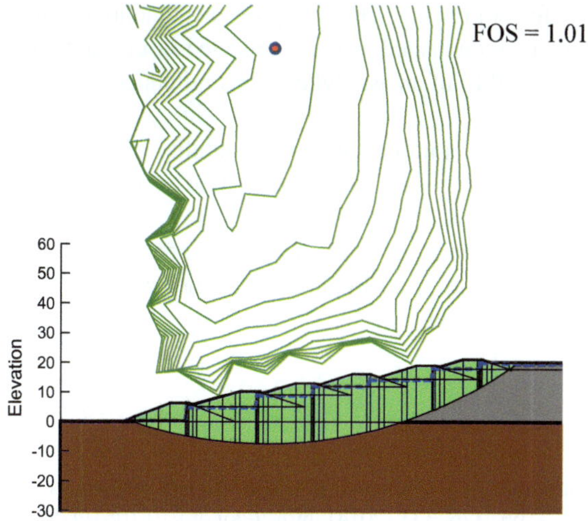

Fig. 10. Critical failure surface after 5th raising in seismic condition (γ_{sat} 15 kN/m^3)

7.3 Discussion

The density of the soil contributes both in driving as well as resisting forces in slope stability analysis. While the normal component of the weight mobilizes shearing resistance along the slip surface, the component parallel to the failure surface acts as the driving force. The shear strength of a material is governed by the effective stress. In lightweight materials like coal ash, even a small increase in density increases the effective stress significantly which contributes in increasing the shearing resistance. Therefore, with an increase in density of hydraulically deposited ash, the ash dyke becomes less susceptible to failure.

In order to study the influence of unit weight of ash on the stability of ash dykes, the angle of shearing resistance was kept constant. The increase in angle of shearing resistance for ash deposited at higher unit weight will increase the factor of safety. Consequently, the percentage difference between FOS at low and high unit weight will increase further.

8 Conclusion

The in situ density of the deposited ash throughout its depth in ash ponds is very difficult to measure precisely due to difficulty in retrieving of undisturbed samples. From literature review, it was observed that that the dry unit weight of loose ash varies from 7.7 to 10.5 kN/m^3. Using appropriate specific gravity of ash, the minimum γ_{sat} of ash varies from 11.7 to 15.8 kN/m^3. It was observed that for γ_{sat} of 11.7 kN/m^3 the FOS was lower by 12.5% in comparison to γ_{sat} of 16 kN/m^3 for 5th raising and by 14% for 6th raising in static condition. Similarly, in seismic condition, the FOS at γ_{sat} of 11.7 kN/m^3 was observed to be less by 14% than that at γ_{sat} of 16 kN/m^3 for 5th raising and by 15% for 6th raising. This study reveals that approximation in estimating the field density of slurry deposited ash gets reflected in the factor of safety of the ash dykes significantly. Stability of an ash dyke is thus influenced by the unit weight of the slurry deposited coal ash.

The role of unit weight in influencing slope stability of incrementally raised dykes of slurry ponds has been reported for the first time in this study and it shows that light-weight deposits such as coal ash can lead to low FOS when the failure surface passes through the slurry-deposited material. Hence, precise estimate of unit weight of the slurry deposited coal ash is an important parameter in safe design of ash dykes raised by the upstream method.

References

Bachus, R.C., Santamarina, J.C.: Geotechnical properties and diagenesis of ponded fly ash. In: Geo-Congress 2014, pp. 326–333 (2014). https://doi.org/10.1061/9780784413272.032

Bjerrum, L., Simons, N.E.: Comparison of shear strength characteristics of normally consolidated clays. Research Conference on Shear Strength of Cohesive Soils, Boulder, pp. 711–726. ASCE, Reston (1960)

Bolton, M.D.: The strength and dilatancy of sands. Géotechnique **36**, 65–78 (1986). https://doi.org/10.1680/geot.1986.36.1.65

Das, S.K., Yudhbir .: Geotechnical characterization of some Indian fly ashes. J. Mater. Civil Eng. **17**, 544–552 (2005). https://doi.org/10.1061/(asce)0899-1561(2005)17:5(544)

Dayal, U.: Raising the Existing Dyke at Korba Super Thermal Power Project using Coal Ash. Design Report No. CN/CE (UD)8826. IIT Kanpur, Kanpur (1989)

Gandhi, S.R., Dey, A.K., Selvam, S.: Densification of pond ash by blasting. J. Geotech. Geoenviron. Eng. **125**, 889–899 (1999)

Gray, D.H., Lin, Y.-K.: Engineering properties of compacted fly ash. J. Soil Mech. Found. Div. **98**(SM4) (1972)

Horiuchi, S., Kawaguchi, M., Yasuhara, K.: Effective use of fly ash slurry as fill material. J. Hazard. Mater. **76**, 301–337 (2000). https://doi.org/10.1016/S0304-3894(00)00205-3

Jakka, R.S., Ramana, G.V., Datta, M.: Shear behaviour of loose and compacted pond ash. Geotech. Geol. Eng. **28**, 763–778 (2010). https://doi.org/10.1007/s10706-010-9337-1

Jakka, R.S., Ramaiah, B.J., Ramana, G.V.: Dynamic characterization of settled pond ash using measured shear wave velocity (Vs) and SPT-N values. Int. J. Geotech. Earthq. Eng. **2**, 83–97 (2011). https://doi.org/10.4018/jgee.2011010106

Kanja, M.A., Wolle, C.M.: Residual strength—new testing and microstructure. In: 9th International Conference on Soil Mechanics and Foundation Engineering, pp. 153–154 (1977)

Kim, B., Prezzi, M., Salgado, R.: Geotechnical properties of fly and bottom ash mixtures for use in highway embankments. J. Geotech. Geoenviron. Eng. **131**, 914–924 (2005)

Ladd, C., Foott, R., Ishihara, K., Schlosser, F., Poulos, H.G.: Stress-deformation and strength characteristics. In: 9th International Conference on Soil Mechanics and Foundation Engineering, pp. 421–494 (1977)

Meyerhof, G.G.: Compaction of sands and the bearing capacity of piles. J. Soil Mech. Found. Div. **85**, 1–29 (1959)

Mishra, M.K., Karanam, U.M.R.: Geotechnical characterization of fly ash composites for backfilling mine voids. Geotech. Geol. Eng. **24**, 1749–1765 (2006). https://doi.org/10.1007/s10706-006-6805-8

Olsen, H.W., Rice, T.L., Mayne, P.W., Singh, R.D.: Piston core properties and disturbance effects. J. Geotech. Eng. **112**, 608–625 (1986)

Pandian, N.S.: Fly ash characterization with reference to geotechnical applications. J. Indian Inst. Sci. **84**, 189–216 (2004)

Pant, A., Datta, M., Ramana, G.V.: Influence of engineering behaviour of coal ash on design of ash dykes. In: Indian Geotechnical Conference, Chennai (2016)

Porbaha, A., Pradhan, T.B.S., Yamane, N.: Time effect on shear strength and permeability of fly ash. J. Energy Eng. **126**, 15–31 (2000)

Seals, R.K., Moulton, L.K., Ruth, B.E. Bottom ash: an engineering material. J. Soil Mech. Found. Div. **98** (1972)

Sridharan, A.: Coal ashes in geotechnical engineering practice: beneficial aspects. In: Emerging Trends in Geotechnical Engineering (2012)

Toth, P.S., Chan, H.T., Cragg, C.B.: Coal ash as structural fill with special reference to Ontario experience. Can. Geotech. J. **25**, 694–704 (1988). https://doi.org/10.1139/t88-080

Impact Assessment of Climate Change in Texas Pavements and Resiliency Strategy

Megha Sharma[1(✉)], Sundeep Inti[2], and Vivek Tandon[1]

[1] The University of Texas at El Paso, El Paso, TX, USA
msharma4@miners.utep.edu, vivek@utep.edu
[2] MixOnSite USA Inc., Buffalo Grove, IL, USA
sundeep@mixonsite.com

Abstract. Changing climate can cause extreme weather events, and variations in hydrologic, storm and temperature patterns. These changes have an adverse impact on the performance of pavement system in Texas. It is of extreme importance to estimate the possible effects of climate change on the performance of the pavements. These estimates will help transportation officials in making decisions to accommodate the climate change into planning and design of pavement systems. The objectives of this study are to document anticipated change in climate and its impact on the performance of the pavements and to present various precautionary measures that can be adopted in current pavement design to withstand the anticipated climate change. In this study, future climate data is selected from North American Regional Climate Change Assessment Program (NARCCAP) databases. Pavement Mechanistic-Empirical (ME) Design software is used for analysis of the pavement performance with historical climate as well as predicted future climate data. Comparing the pavement performance with historical climate and predicted future climate data revealed that the distresses in the pavement structure is increasing with the future climate scenario. Various adaptation approaches of pavement design to survive the future climate change are recommended through this study.

1 Introduction

Intergovernmental Policy on Climate Change (IPCC) has identified that the climate pattern has gradually changed in the last century. The temperature of the atmosphere has been steadily rising, the area covering snow has gradually decreased, sea level has been rising, among others. The major factor contributing to climate change is a significant increase in greenhouse gas (GHG) emissions due to anthropogenic activities (industrialization and a significant increase in global population). Changing climate can cause extreme weather events, and variations in hydrologic, storm and temperature patterns, which can be detrimental to the world economy. Various national and international agencies have spent significant resources on predicting future climates like United States Geological Survey (USGS), National Oceanic and Atmospheric Administration (NOAA), Geophysical Fluid Dynamics Laboratory (GFDL) to name the few. These agencies have used different emission scenarios to forecast the future climate that can be used to develop strategies to mitigate the influence of climate changing.

M.-C. Weng et al. (eds.), *Current Geotechnical Engineering Aspects of Civil Infrastructures*, Sustainable Civil Infrastructures,
https://doi.org/10.1007/978-3-319-95750-0_4

The economy of a country significantly depends on the transportation infrastructure and climate change can adversely impact the performance of transportation infrastructure because it is continuously exposed to the climate. Studies have been carried out to study the vulnerability of the transportation network to climate change.

United States Department of Transportation (US DOT) Gulf Coast Study (Phase 1), conducted vulnerability study and observed that the rise in sea levels would affect the interstates and arterials, 75% of the port facilities, and the average temperature could increase. As per Sultana et al. (2014), the strength of the pavements reduces by 1.5–50% due to flooding. Meagher et al. (2012) used Mechanistic-Empirical Pavement Design Guide (MEPDG) to predict pavement distress by transforming the climate data with future temperature change, observing an impact on AC rutting with negligible impact on alligator cracking.

In a joint project (ROADAPT research project) funded by European countries, many countries have developed climate adaptation strategies, and policies like the United Kingdom requires to consider effects of climate change in planning and design stages. Similarly, New Zealand government provides information and guidance to the local authorities on integrating climate change considerations into decision-making processes.

Although climate influences all modes of the transportation, it is beyond the focus of this study. Therefore, this study is mainly focused on pavement infrastructure. The pavement design uses historical climate data to estimate service life of the designed pavement. Since the pavement infrastructure will be exposed to the changing climate, it is very likely that the estimated service life will be reduced and significant economical and natural resources will be required to maintain and rehabilitate pavement infrastructure. To mitigate the influence of climate change, the transportation agencies need to use future climate data in designing future pavement infrastructure that is resilient to climate change. Thus, the main objectives of this study are:

(a) Evaluate influence of climate change parameters on the performance of pavements, and;
(b) Evaluate adaptation methods to enhance the resiliency of pavement infrastructure.

Pavement Mechanistic-Empirical (ME) Design software for predicting pavement performance was used for estimating design service life of pavements. The Pavement ME design software uses Integrated Climate Model (ICM) to account for climate influence. ICM consists of weather details (Temperature, Wind Speed, Relative Humidity, Sunshine, and Precipitation) for the geographical location of the project at an hourly interval. Although software consists of historical weather database and doesn't take into account the changing climate, North American Regional Climate Change Assessment Program (NARCCAP) climate data source was extracted and was used for estimating design service life.

2 Climate Models

NARCCAP utilizes Special Report on Emission Scenario (SRES) A2 emission scenarios for predicting future climate. This climate data source uses a dynamic downscaling method to produce the data at the local resolution of 31 × 31 miles (50 × 50 km). NARCCAP includes six Regional Climate Models (RCMs) and four Global Climate Models (GCMs) namely: RCMs are (1) Hadley Regional Climate Model Version 3 (HRM3), (2) Regional Climate Model 3.0 (RCM3), (3) the Canadian Regional Climate Model (CRCM), (4) Experimental Climate Predication Center Regional Spectral Model (ECPC), (5) Mesoscale Meteorological Model Version 5.0 (MM5I), and (6) the Weather Research and Forecasting Model (WRFG), GCMs are (1) the Hadley Centre Climate Model (HadCM3), (2) Community Climate System Model (CCSM), (3) the Canadian Global Climate Model (CGCM3), and (4) the Geophysical Fluid Dynamics Laboratory (GFDL) model.

The eleven climate model obtained from NARCCAP is from here onwards referred as CRCM-CCSM, CRCM-CGCM3, HRM3-HADCM3, HRM3-GFDL, MM5I-HADCM3, MM5I-CCSM, RCM3-CGCM3, RCM3-GFDL, WRFG-CCSM, WRFG-CGCM3, and ECP2-HADCM3, (RCMs-GCMs, i.e., CRCM is RCM, and CCSM is GCM in CRCM-CCSM model).

NARCCAP climate data simulations are available for a current period of 1968–2000 and future simulation from 2038 to 2070.

3 Methodology

To evaluate the influence of the climate change on pavement service life, an existing Fort Worth (Texas) pavement design was selected. The thickness of individual layers is shown in Fig. 1, and relevant traffic information and location of the site is included in Table 1. The design analysis is performed for 20 years of service life. Although pavement design software predicts various performance parameters, this study evaluated rutting and International Roughness Index (IRI) while analyzing the performance of the pavement section. Based on the literature review and discussions with TxDOT engineers, it was identified that maintenance measures are taken when IRI exceeds 100 in./mile and rutting is more than 0.4 in.

The following steps outline performance prediction process:

1. North American Regional Climate Change Assessment Program (NARCCAP) climate data source was used for generating climate data files to be used in Pavement ME Design software using the method described in Meagher et al. (2012) and Sharma et al. (2017).
2. The influence of the following weather condition changes on the pavement performance:
   (a) Using the historical climate database for analysis.
   (b) Using the future climate parameters (mainly temperature and precipitation) obtained from NARCCAP.

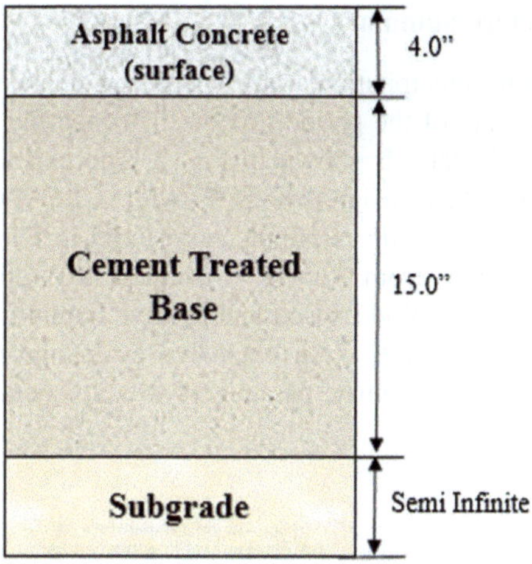

Fig. 1. IH 30 frontage road pavement section selected for this study

Table 1. Design criteria for pavement section in fort worth, TX

Highway location	IH 30 Frontage road, Tarrant County
Annual average daily traffic (AADT) beginning	22,990
Annual average daily truck traffic (AADTT)/percentage of trucks	828/3.6%
Analysis period	20 years

   (c) Changing the water table depth to consider a change in bearing capacity of the subgrade layer due to saturation.
   (d) Considering both changes in climate parameters and bearing capacity of the subgrade layer.
3. Considering adaptation method:
   (a) Increasing the thickness of the Asphalt Concrete (AC) layer.
   (b) Changing the Performance Grade (PG) binder.
   (c) Changing AC layer mix type.
   (d) Changing the modulus of the subgrade layer.

4 Results and Discussion

The results of the analysis are discussed in the following paragraphs for the pavement section analyzed (Fig. 1).

4.1 Climate Parameters Summary

Mean annual temperature, precipitation, wind speed and relative humidity for eleven climate models for 30 years of the period (2038–2070) and the existing climate conditions (Pavement ME Climate) is shown in Fig. 2. The existing mean annual temperature used in Pavement ME Design is 66.8 °F, while CRCM-CCSM climate model estimates that the mean annual temperature increases to 69.1 °F in the future. The mean annual precipitation increases from 33.9 in. (Pavement ME Climate) to 59.7 in. for CRCM-CCSM climate model. Wind speed and relative humidity increase from 9.8 to 13.9 mph and 65–71% respectively. An extensive evaluation (Sharma et al. 2017) identified that most influencing climate parameters were temperature and precipitation and were used in the analysis.

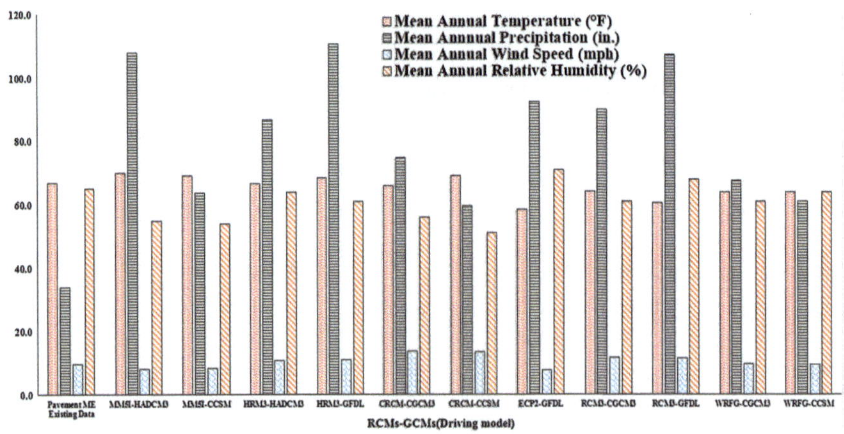

Fig. 2. Historical and future climate data for fort worth

4.2 Pavement Distresses

To minimize clutter, the data extracted from CRCM-CCSM climate model is used for explaining the results because the predictions from the selected climate model were closer to the average predictions from all the future climate models. The rutting and roughness index estimated from the analysis are shown in Figs. 3 and 4, respectively.

4.2.1 International Roughness Index (IRI)

The estimated IRI shown in Fig. 3 suggests that the IRI is dependent on the climate. An IRI value of 100 in. /mile (a milestone for triggering maintenance) will be attained within 9.6 years of service (with historical climate data) while same IRI will be attained within 8.3 years of service (with CRCM-CCSM data). Thus, maintenance will be scheduled a year earlier than planned.

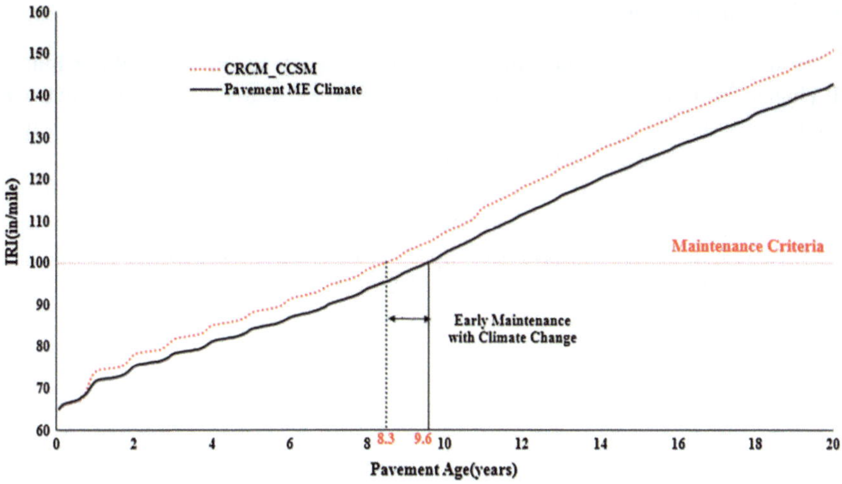

Fig. 3. Influence of climate change on IRI

4.2.2 Asphalt Concrete (AC) Rutting

Figure 4 illustrates the change in AC Rutting for 20 years based on historical weather data (Pavement ME Climate) and future weather data (CRCM-CCSM model). Based on the maintenance criterion (an action is needed if the pavement section reaches 0.4 in. of AC rutting), the maintenance is required after 6 years of service (CRCM-CCSM data) while maintenance is required after 15.8 years of service (historical climate). This suggests that change in climate will reduce service life by 9 plus years requiring some rehabilitation activity earlier than anticipated.

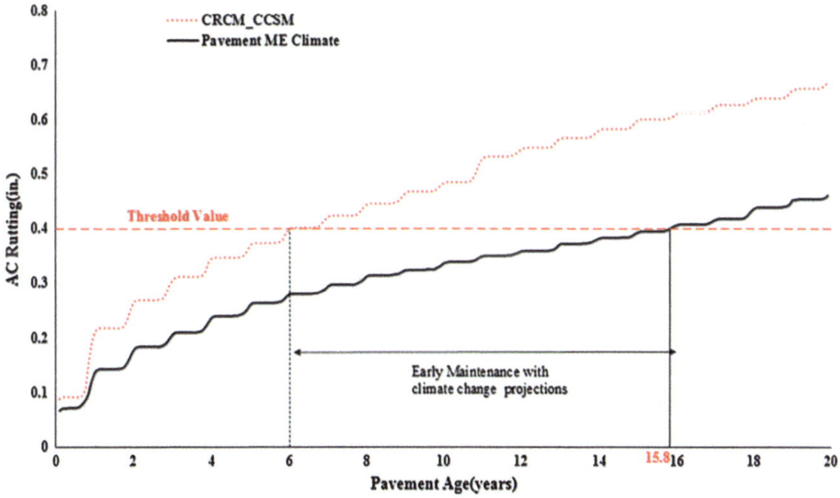

Fig. 4. Influence of climate on AC rutting

4.3 Saturation Damage

One of the predictions of climate models is that frequency and intensity of precipitation will change in future. In the case of heavy rainfall, the water table will increase and saturate the subgrade layer. The bearing capacity of subgrade layer will reduce under saturation and is evaluated in this study. It is assumed that extreme rainfall is occurring at an interval of 1, 2, 3, 5, 10, and 15 years of service of the pavements, saturating the subgrade soil for 7.5 days, 15 days, one month and two months. During this heavy rainfall period, the water table is raised closer to the subgrade soil. Since the focus was to evaluate the influence of saturation, the historical climate data files were used in the analyses. The influence of the change in water table depth on the pavement distresses is summarized in Figs. 5 through 7.

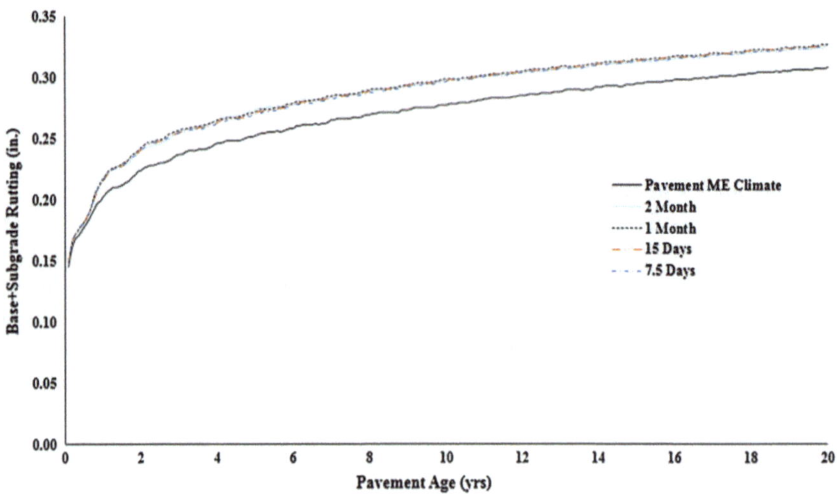

Fig. 5. Combined base and subgrade rutting

Due to saturation of the subgrade, there is no variation in AC rutting; the only effect will be on the subgrade rutting. Since the software generates AC and total rutting, the AC rutting was subtracted from the total rutting and is shown in Fig. 5. The trend shows an increase in rutting when subgrade is saturated. The influence of rutting on IRI is shown in Fig. 6.

The Fig. 7 changes in subgrade modulus due to saturation. The trend indicates that modulus drops with saturation, but after lowering of water table, the modulus goes back to previous levels. The drop in modulus suggests that the bearing capacity of subgrade layer is reduced, but the IRI change is not significant. In this case, it can be concluded that the influence of saturation is minimal; however, this was not true for other parts of the state where the soil is moisture dependent. The saturation of subgrade layer reduces the bearing capacity of soils, especially in Eastern Texas. The loss of modulus and recovery of the modulus after reduction in saturation is as expected. However, permanent loss in the modulus needs to be further evaluated to identify the

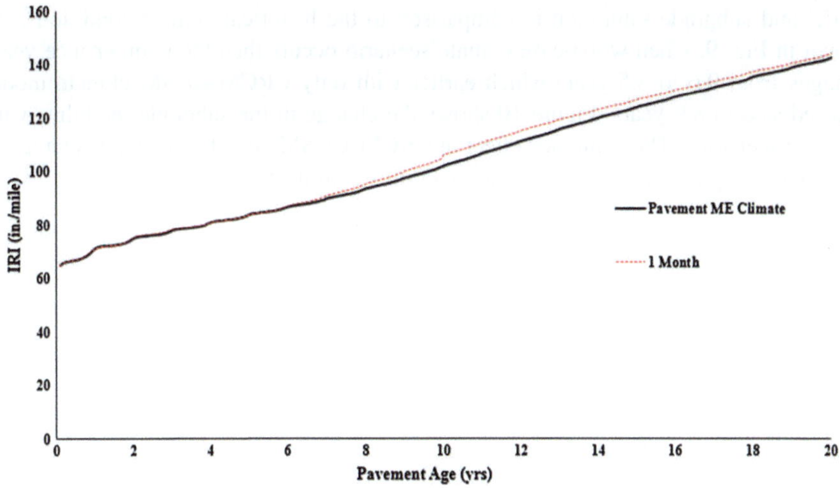

Fig. 6. Influence of subgrade saturation on IRI of the pavement section

reasons for the loss. If true, the level of reduction in the modulus suggests that with an increase in truck traffic may further enhance damage because of loss of modulus (factor of safety).

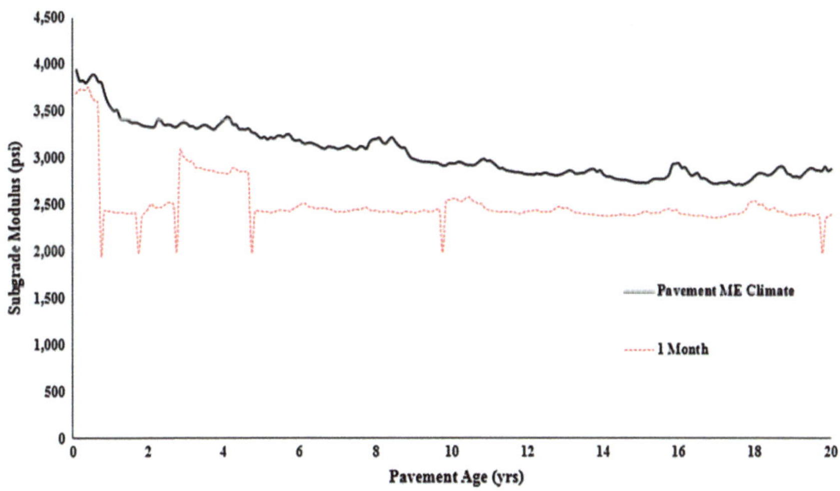

Fig. 7. Change in subgrade modulus due to saturation

4.4 Moisture Damage and Climate Change Parameters

A worst-case scenario was considered when the simulated future climate files are used along with the subgrade moisture damage (changing the water table depth). Figures 8, 9 and 10 shows the changes in the distress with the inclusion of CRCM-CCSM climate model and subgrade saturation in comparison to the historical climate conditions. As shown in Fig. 9, when worst-case climate scenario occurs then the maintenance years changes from 9.6 to 8.5 years which earlier with only CRCM-CCSM climate model was reduced to 8.9 years. Figure 10 shows the change in the subgrade modulus, with different scenarios. The combined effect of CRCM-CCSM climate model and change in water table depth reduces the subgrade modulus significantly.

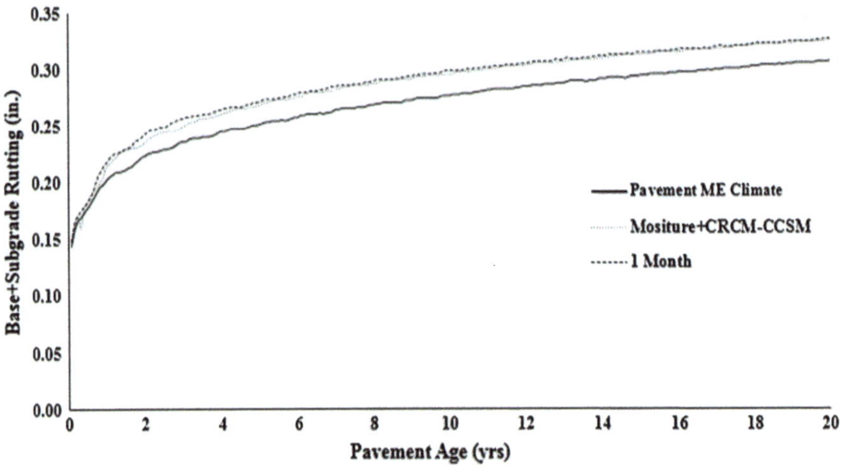

Fig. 8. Combined base and subgrade rutting of the pavement section

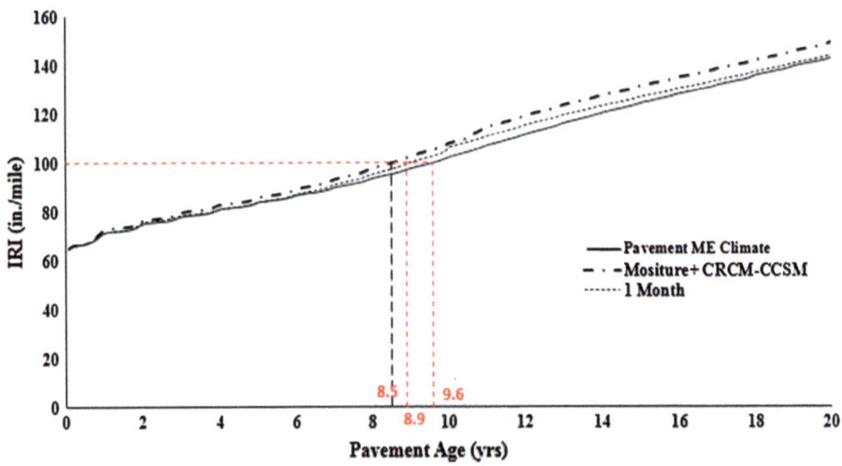

Fig. 9. Influence of climate model and saturation on IRI of the pavement section

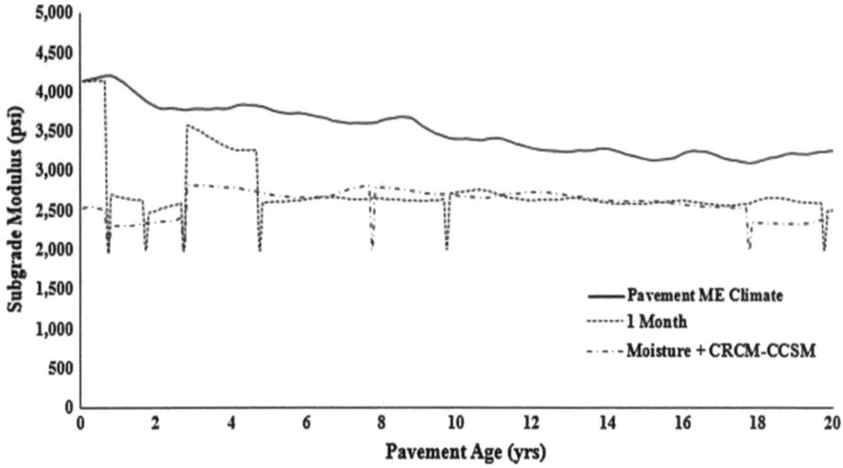

Fig. 10. Influence of climate model and saturation on subgrade modulus of the pavement section

4.5 Adaptation Methods

One of the objectives of this study was to propose adaptation strategies for mitigating the influence of climate. Since pavement section is located in a regional location, the adaptation strategies need to be regional because pavement sections hundreds of miles apart need different strategies. To withstand changing climate either pavements need to use high-performance materials or enhanced layer thickness or both. The following strategies to minimize the influence of climate change, have been proposed that can enhance the resiliency of the pavements, explained in the subsequent sections.

4.5.1 Adaptation Methods for Climate Change

1. Increasing the Thickness of AC Layer

Adaptation is required to maintain the resiliency of the pavement sections in the event of climate change. One of the adaptation strategies can be to increase the thickness of the AC layer. For the pavement section in Fort Worth (Texas), the thickness of the AC layer is increased by an inch, and the performance of the modified pavement section is similar to the one with the historical climate data. As we can see in Fig. 11, the increase in the thickness of the AC layer improves the performance of pavement section (IRI). In the figure, CC refers to climate change for CRCM-CCSM climate prediction models. With existing climate and design, the section was required to have maintenance after 9.1 years while if the climate change occurs the performance of the section will decrease, and maintenance is required after 8.3 years. If we consider the climate change early on in design and construction, the pavement section will require maintenance after 9.4 years similar to the existing conditions.

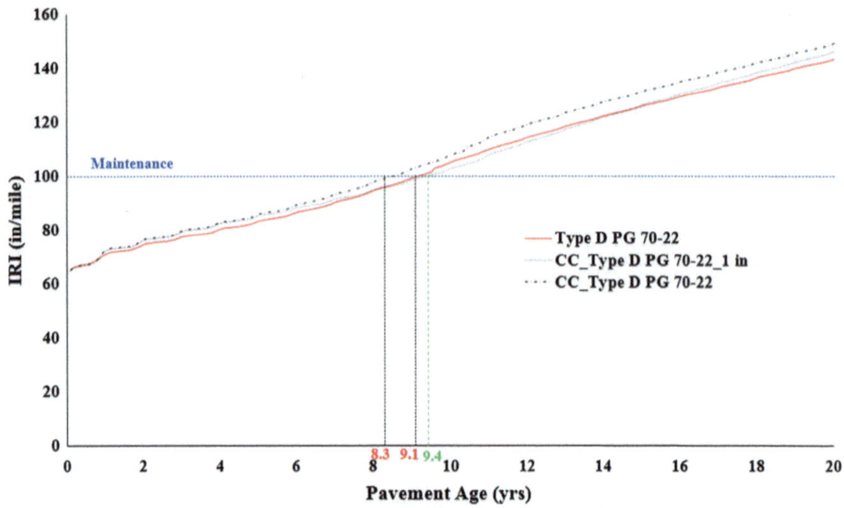

Fig. 11. Influence of AC thickness on pavement performance

2. Binder Grade Change

Changing the Performance Grade (PG) of the binder can also improve the performance of the pavement section. Figure 12 shows the change in IRI of the pavement sections for different binder grade types. If we change the binder grade from PG 70-22 to PG 76-22 for CRCM-CCSM climate model the maintenance year change from 8.3 to 8.8 years.

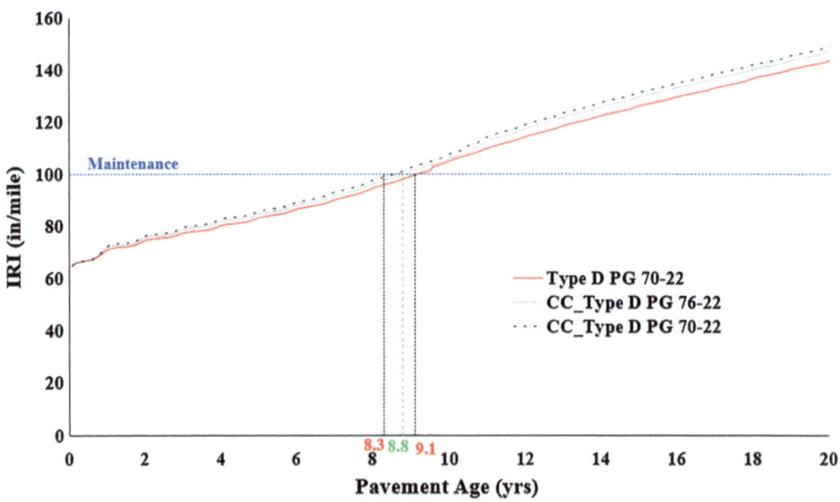

Fig. 12. Influence of binder grade type on pavement performance

3. Increasing the Thickness of AC Layer and Changing Binder Grade

Changing the Performance Grade (PG) along with layer thickness further improves the performance of the pavement sections as well. If the binder grade is changed from PG 70-22 to PG 76-22 and the increase in thickness by one inch for CRCM-CCSM climate model, the maintenance year change from 8.3 to 9.8 years (Fig. 13).

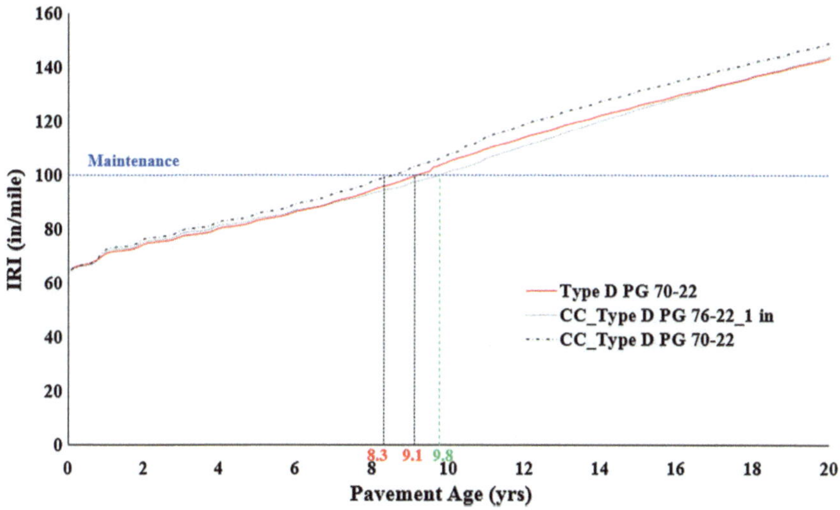

Fig. 13. Change in performance of pavements with binder grade change and increase in AC thickness

4. Changing Mix Type

Similarly, the third adaptation strategy could be to change the mix type of the AC layer. Table 2 shows the AC mix types typically used along with the reference for the mix specifications used by TxDOT. Figures 14 and 15 illustrates AC Rutting, and IRI distresses of the selected pavement section over the years with the change in the mix types for the AC layer. Type C and Type D behave similarly while using SP mix will improve the performance of the pavement section.

Table 2. Asphalt mix type specifications

Item type	Specifications
Dense-graded hot mix asphalt (Item 340, 341) Type C, Type D	Texas Department of Transportation (TxDOT) "Standard Specifications for Construction and Maintenance of Highways, Streets, and Bridges"
Stone Matrix Asphalt (SMA) (Item 346)	
Superpave Mix- Coarse (SP-C) or Performance Mix (Item 344)	

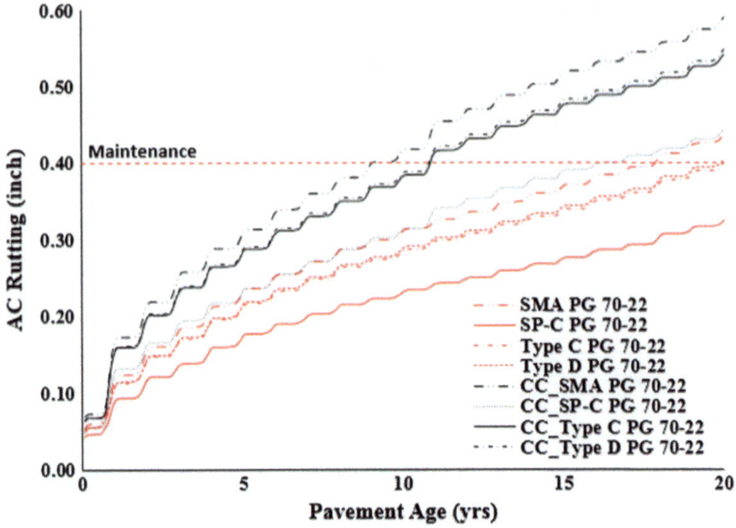

Fig. 14. AC rutting of the pavement section for different AC mix types

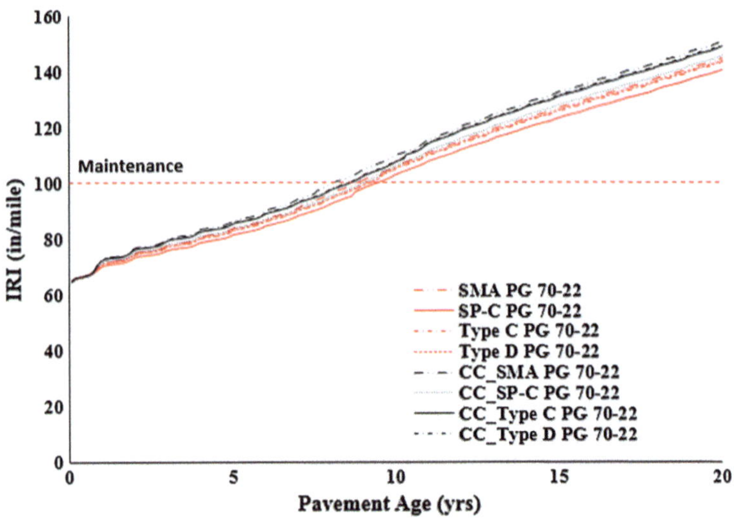

Fig. 15. IRI of the pavement section for different AC mix types

4.5.2 Adaptation Methods for Moisture Damage

Improving Subgrade Modulus: Heavy rainfall influences the water table depth, making the sub-grade completely saturated during heavy precipitation season. One of the adaptation methods could be to improve the subgrade material quality against the moisture damage. In Fig. 16, it can be observed that by enhancing sub-grade (higher modulus) from 4.5 ksi (initial modulus) to 6 ksi, there is an improvement in the performance of the pavements.

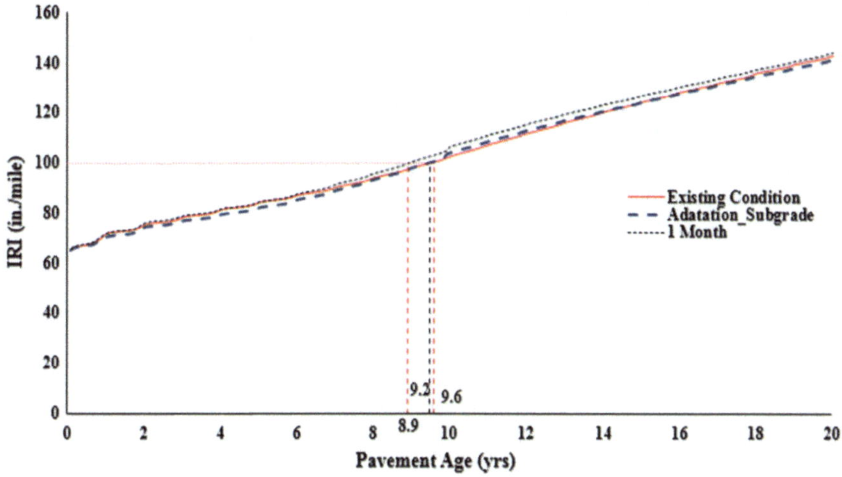

Fig. 16. Adaptation to moisture damage in the subgrade

4.5.3 Adaptation to Moisture Damage and Climate Change

In the case of extreme events (considering changing climate along with moisture damage) the adaptation is shown in Fig. 17. The increase in the thickness of the AC layer along with improving the sub-grade quality improves the performance of the pavement sections leading to resiliency in the event of a change in climate.

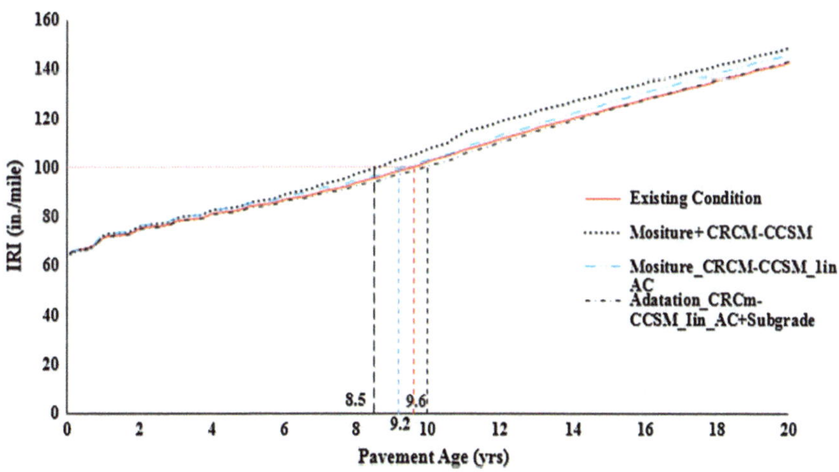

Fig. 17. Adaptation to moisture damage and climate change

5 Closure

- Eleven climate models were considered in this study, and every model simulated the different change in mean annual temperature and precipitation. So, average model simulation is used for analyzing the pavement performance.
- The service life of pavements is reduced in the event of change climate and extreme events.
- Increase in AC layer helps in mitigating the impact of the climate change.
- Change in the PG grade and/or use of better performing mix types enhances the performance of the pavement sections in the event of climate change.
- Better quality of the subgrade (high modulus) maintains the pavement performance if extreme event leads to a reduction in bearing capacity of the subgrade layer.

Acknowledgements. The authors like to acknowledge Texas Department of Transportation (TxDOT) for financial support for this project.

References

IPCC—Intergovernmental Panel on Climate Change. http://www.ipcc.ch/ (2016). Accessed 26 July 2016

Intergovernmental Panel on Climate Change: Working Group III. Emissions Scenarios: Summary for Policymakers. A Special Report of IPCC Working Group III. Intergovernmental Panel on Climate Change (2000)

Stocker, T.F., et al.: Climate change 2013: The physical science basis. Cambridge University Press, New York (2014)

National Research Council (US): Committee on Climate Change, US Transportation, National Research Council (US). Division on Earth, and Life Studies. Potential Impacts of Climate Change on US Transportation, vol. 290. Transportation Research Board (2008)

Pachauri, R.K., Reisinger, A.: Contribution of Working Groups I, II and III to the Fourth Assessment Report of the Intergovernmental Panel on Climate Change. IPCC, Geneva, Switzerland, p. 104 (2007)

Savonis, M.J., et al.: Impacts of Climate Change and Variability on Transportation Systems and Infrastructure: Gulf Coast Study, Phase I. U.S. Climate Change Science Program, and Global Change Research (2008)

Sultana, M., et al.: A review of the structural performance of flooded pavements. In: 26th ARRB Conference—Research Driving Efficiency (2014)

Meagher, W., et al.: Method for evaluating implications of climate change for design and performance of flexible pavements. Transp. Res. Rec. J. Transp. Res. Board 2305, 111–120 (2012). Transportation Research Board of the National Academies, Washington, DC

Filosa, G., Oster, A.: International Practices on Climate Adaptation in Transportation. Publication DOT-VNTSC-FHWA-15-01 (2015)

Guide for Mechanistic-Empirical Design of new and rehabilitated pavement structures. NCHRP Project 1-37A (2004)

Mearns, L.O., et al.: A regional climate change assessment program for North America. EOS **90** (36), 311–312 (2009)

Mearns, L.O., et al.: The North American Regional Climate Change Assessment Program dataset. National Center for Atmospheric Research Earth System Grid data portal, Boulder, CO. Updated 2014, Data downloaded 2015-10-23 (2007)
Sharma, M., et al.: Influence of climate prediction models on pavement performance. In: Transportation Research Board 96th Annual Meeting, No. 17-05944 (2017)

Implementation of Computational Contact Analysis in Geotechnical Engineering

Daqing Xu[1(✉)], Luis G. Vasquez[2], and William M. Isenhower[1(✉)]

[1] Ensoft, Inc, 3003 West Howard Lane, Austin 78728, TX, USA
(512) 244-6464lcra@ensoftinc.com
[2] Ensoft, Inc./Lymon C. Reese and Associates, Inc, 3003 West Howard Lane, Austin 78728, TX, USA

Abstract. Analyses of mechanical interactions between soil and structural members in geotechnical engineering are long-standing challenges in the design of piles, foundations, retaining walls, and culverts. Due to the high nonlinearity of these problems, numerical solutions using finite element method (FEM) and finite differences are dominant in engineering practice. In this paper, a detailed procedure of FEM implementation is presented to solve 2D frictional interaction problems with finite sliding. This presentation starts from the geometrically exact theory to express the kinematics of contact, virtual work equation and linearization in the local convective coordinate system. Formulations of the contact tangent stiffness become straightforward and easy to understand. Tangential behavior is simulated by the regularized Coulomb friction law and constraints are enforced by the penalty approach. If linear finite elements are used, local smoothing is essential to improve convergence performance. Two applications in geotechnical engineering are presented to demonstrate the capabilities of this implementation.

Keywords: Node-to-segment · Penalty · Frictional · Finite element
Covariant coordinate

1 Introduction

Soil-structure interaction problems are frequently encountered in geotechnical engineering, such as interactions between soil-culvert, soil-retaining wall, soil-pile, soil-concrete dam, etc. Contact states can be "in contact" or "open" in the normal direction, and "sticking" or "sliding" in the tangential direction. These problems are highly nonlinear in nature and closed-form solutions are only available for few cases. Most solutions have to be obtained by numerical approaches like the finite element method (FEM). After more than thirty years of effort, a framework for analysis of contact problems has been well established and general-purpose FEM programs are available to solve complicated contact problems in today's engineering practice. However, due to complexity in theory and significant difficulties in the development of programs, availability of FEM software to solve interaction problems in geotechnical engineering is still limited.

M.-C. Weng et al. (eds.), *Current Geotechnical Engineering Aspects of Civil Infrastructures*, Sustainable Civil Infrastructures,
https://doi.org/10.1007/978-3-319-95750-0_5

Using FEM to solve soil-structure interactions can trace its origin to the early 1960s. Goodman [1] developed an interface element with zero thickness to simulate the interface behavior in rock. Desai and Zaman [2] proposed a thin-layer interface element to simulate rock joints. In these methods, node-to-node contact status must be maintained. This approach has the benefit of convenient implementation. However, when the interface has finite sliding this approach can have major difficulties because node-to-node contact status cannot be maintained.

In solid mechanics, versatile approaches have been developed to solve general multi-body contact problems. References [3–5] are early advances in this field where Lagrange multiplier method or the penalty method is employed to apply contact constraints. The interface can be frictionless or frictional. Among various approaches, the penalty method has attracted more interest. Although impenetrability (non-penetration) is only approximately satisfied, the penalty approach has advantages of good convergence performance and convenient incorporation into current FE codes. Kikuchi and Oden [6] have discussed theoretical formulations and implementation of penalty approach for frictionless and frictional contacts. Other early references related with the penalty approach can be found in [7, 8].

In linearization of the variational equation, the *direct approach* is popular in publications in which linearization is applied to discrete configurations. However, procedures of linearization to generate a consistent stiffness matrix are usually complicated and difficult to understand. An alternative approach is known as *covariant approach*. Kinematics of contact bodies are expressed in local coordinate system. Linearization is applied to the weak form of the continuous version to obtain the tangent contact stiffness in a covariant form. This approach involves many covariant operations; however, formulations of contact stiffness are straightforward and more understandable. Details of this approach can be found in [9, 10].

In this work, implementation of computational contact analysis to solve problems in geotechnical engineering is introduced. After the introduction section, the kinematics of contact, virtual work and tangent stiffness in the covariant form are described. Following is the section about finite element discretization, calculation of closest point projection (CPP) and other issues related with program development. In the last section, two applications in geotechnical engineering are presented.

2 Kinematics of Contact

The configuration of a contact problem between two solid bodies is depicted in Fig. 1. We use the symbols Ω^A and Ω^B to denote solid bodies and Γ_u, Γ_σ, and Γ_c to denote the displacement boundary, force boundary and contact boundary respectively. The terminology "master-slave" is adopted to distinguish Ω^A from Ω^B. The "master" body has a larger Young's modulus or stiffer material. In geotechnical engineering, structures are usually concrete or steel. Thus, the "master" title is assigned to the structure and "slave" is assigned to the soil generally. In the "node-to-segment" approach discussed in this work, the terminology "contact pairs" is used to describe the interaction on the interface. Each contact pair includes one "slave" point from the slave body and one piece of the "master" 2D curve or 3D surface as depicted in Fig. 2. The local coordinate system is defined on the master body side. We denote the position vector of the slave point by $\boldsymbol{r}_s$ and the master surface by the position vector

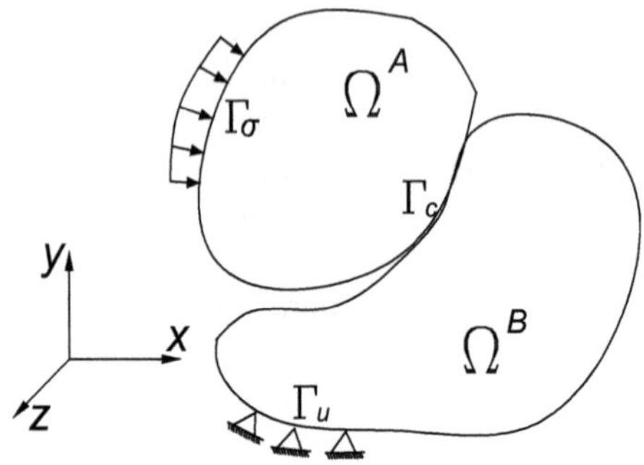

Fig. 1. Configuration of two-body contact problem

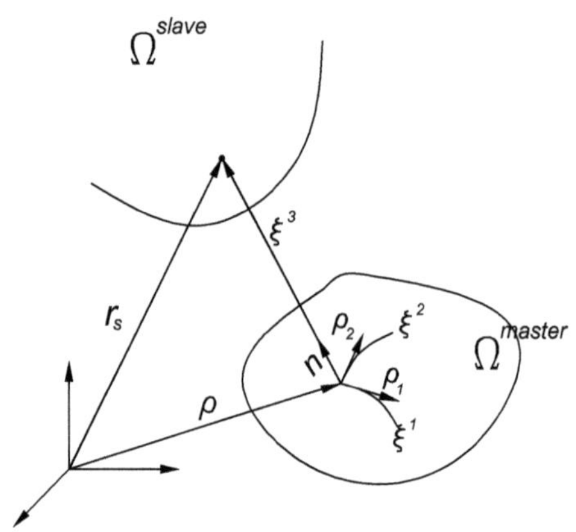

Fig. 2. Local coordinate system and closest point projection

$$\boldsymbol{\rho} = \boldsymbol{\rho}(\xi^1, \xi^2) \tag{1}$$

where ξ^1, ξ^2 are local coordinates. When contact happens, impenetrability should be satisfied so that the slave point has zero gap to its projection on the master surface. The procedure to find the "Closest Point Projection (CPP)" can be expressed as

$$\min||\boldsymbol{r}_s - \boldsymbol{\rho}(\xi^1, \xi^2)|| \tag{2}$$

The normal gap ξ^3 between the slave point and its CPP is computed using

$$\boldsymbol{r}_s = \boldsymbol{\rho}(\xi^1, \xi^2) + \xi^3\mathbf{n} \tag{3}$$

where $\mathbf{n}$ is the outward unit norm on the master surface.

3 Normal Contact

In the normal direction, a slave point may have a positive gap (separation) with zero contact traction, or touch the master surface with zero gap and non-zero contact traction. This geometrical constraint can be described by the Kuhn-Tucker condition by measurement of ξ^3 (see Fig. 2 for definition of ξ^3).

$$\begin{cases} \xi^3 > 0, & \textit{no contact and } t_N = 0 \\ \xi^3 \leq 0, & \textit{in contact and } t_N = \epsilon_N \xi^3 \end{cases} \tag{4}$$

where t_N is the normal contact traction and ϵ_N is the normal penalty parameter. One may note that t_N is negative in Eq. 4 and penetration does not vanish when contact occurs. Although the penalty approach satisfies the non-penetration constraints approximately, results with good engineering accuracy are still available when appropriate penalty parameters are used. In theory, ϵ_N is recommended to be set as large as possible.

4 Tangential Contact

A frictional contact is more common than a frictionless one. The interface of soil-structure interactions is often rough and tangential friction stresses may play an important role in analysis. However, interface friction is a complicated phenomenon that involves normal pressure, roughness of the interface, temperature and relative velocity, etc. Among various approaches, the Coulomb friction law is widely adopted. Classical Coulomb law assumes that there is no relative tangent displacement when the tangential force is smaller than a threshold value. Once sliding starts, the tangent friction force is linearly proportional to the normal pressure. The classical Coulomb law has implementation difficulties due to the sudden change from zero relative displacement to sliding. Regularized Coulomb friction law is recommended which allows a small relative motion before sliding.

Tangential contact status should be distinguished as "sticking" or "sliding". As an analogy to elasto-plasticity, we define the tangential sliding function as

$$\Phi = |\boldsymbol{t}_T| - \mu t_N \tag{5}$$

where μ is the coefficient of interface friction and $\boldsymbol{t}_T$ is the tangential contact traction. Equation 5 has the difficulty of differentiability and regularization is necessary.

We follow the same regularization as in [14], and rewrite the relative sliding velocity $\mathbf{v}_r$ in the regularized form:

$$\frac{dt_T}{dt} = -\epsilon_T \mathbf{v}_r \tag{6}$$

where ϵ_T is the tangential penalty parameter.

The left part in Eq. 6 is the time derivative of the tangent contact tractions. For static problems in 2D assumptions, local coordinates reduce to one parameter and we can use an increment of displacement $\Delta\xi^1$ to replace the rate form of Eq. 6. Thus,

$$T^1_{n+1} - T^1_n = -\epsilon_T \Delta\xi^1 = -\epsilon_T\left(\xi^1_{n+1} - \xi^1_n\right) \tag{7}$$

The subscript n and $n + 1$ represent consecutive load increments. At initial conditions, $T^1_0 = 0$, $\xi^1_0 = \xi_c$, where ξ_c is the initial Closest Point Projection.

Determination of tangential contact status of a slave point under the $n + 1$ load increment has the following steps:

- Calculate trial of $\boldsymbol{t}_T$. In case of a 2D problem, trial of T^1 at $n + 1$ load increment is calculated as

$$T^{1,trial}_{n+1} = -\epsilon_T \Delta\xi^1 \tag{8}$$

- Real tangential traction is determined by the returning mapping algorithm.

$$T^1 = \begin{cases} T^{1,trial}_{n+1} \; if \left|T^{1,trial}_{n+1}\right| \leq \frac{\mu|t_N|}{|\boldsymbol{\rho}_1|} & keepsticking \\ -\frac{\Delta\xi^1}{\left|\Delta\xi^1\right|}\frac{\mu|t_N|}{|\boldsymbol{\rho}_1|} \; if \left|T^{1,trial}_{n+1}\right| \geq \frac{\mu|t_N|}{|\boldsymbol{\rho}_1|} & changetosliding \end{cases} \tag{9}$$

If sliding occurs, update closest point projection ξ_c using

$$\xi_c = \xi^1_{n+1} - \frac{\Delta\xi^1}{\left|\Delta\xi^1\right|}\frac{\mu|t_N|}{\epsilon_T|\boldsymbol{\rho}_1|} \tag{10}$$

where $\boldsymbol{\rho}_1$ is the first derivative $\boldsymbol{\rho}_1 = \frac{\partial \boldsymbol{\rho}}{\partial \xi^1}$

5 Weak Equilibrium

On the contact interface, active contact tractions between Ω^A and Ω^B have the same value, but directions are inverted. Therefore, it is possible to write the total contact traction $\boldsymbol{t}_c$ acting on the slave point as

$$\boldsymbol{t}_c = t_N \mathbf{n} + \boldsymbol{t}_T = t_N \mathbf{n} + T^i \boldsymbol{\rho}_i \tag{11}$$

Assuming the interface deformation to be small and problems to be static, the virtual work of the two-body contact problem has the following form

$$\begin{aligned} &\int_{\Omega^A} \left[\boldsymbol{\sigma}^A : \nabla \delta \boldsymbol{u} - \boldsymbol{b}^A \cdot \delta \boldsymbol{u}\right] d\Omega + \int_{\Omega^B} \left[\boldsymbol{\sigma}^B : \nabla \delta \boldsymbol{u} - \boldsymbol{b}^B \cdot \delta \boldsymbol{u}\right] d\Omega \\ &- \int_{\Gamma_\sigma} \boldsymbol{t}_\sigma^A \cdot \delta \boldsymbol{u} d\Gamma_\sigma - \int_{\Gamma_\sigma} \boldsymbol{t}_\sigma^B \cdot \delta \boldsymbol{u} d\Gamma_\sigma - \int_{\Gamma_c} \boldsymbol{t}_c \cdot (\delta \boldsymbol{r}_s - \delta \boldsymbol{\rho}) d\Gamma_c \end{aligned} \tag{12}$$

where $\delta \boldsymbol{u}$ is the virtual displacement and $\boldsymbol{b}$ is the body force.

The last component in Eq. 12 is the virtual work δW^c contributed from the contact tractions. In 2D conditions, δW^c can be simplified to follows:

$$\delta W^c = \delta W_N^c + \delta W_T^c = \int \left[t_N \delta \xi^3 + T^1 (\boldsymbol{\rho}_1 \cdot \boldsymbol{\rho}_1) \delta \xi^1\right] d\Gamma_c \tag{13}$$

6 Linearization

In this section, we focus on linearization of the contact virtual work δW^c only. Since contact analysis is highly nonlinear, an iterative method such as Newton's method is required. Therefore, it is necessary to develop the tangent stiffness of Eq. 13. Operations of the time derivative to Eq. 13 are quite complicated. Thus, this part is omitted in this paper. Interested readers can refer to [9–11] for more detail. Here we only list results after linearization. Assuming the problems are 2D, the normal and tangential contact stiffness have the following formulations:

7 Directional Derivative of the Normal Part

$$\begin{aligned} D_v\left(\delta W_N^c\right) = &\int_s \epsilon_N H\left(-\xi^3\right)(\delta \boldsymbol{r}_s - \delta \boldsymbol{\rho}) \cdot (\mathbf{n} \otimes \mathbf{n})(\mathbf{v}_s - \mathbf{v}) ds \\ &- \int_s \epsilon_N \xi^3 H\left(-\xi^3\right) \left[\delta \boldsymbol{\tau} \cdot (\mathbf{n} \otimes \boldsymbol{\tau})(\mathbf{v}_s - \mathbf{v}) + (\delta \boldsymbol{r}_s - \delta \boldsymbol{\rho}) \cdot (\boldsymbol{\tau} \otimes \mathbf{n}) \frac{\partial \boldsymbol{\tau}}{\partial t}\right] ds \\ &- \int_s \epsilon_N \xi^3 H\left(-\xi^3\right) \kappa (\delta \boldsymbol{r}_s - \delta \boldsymbol{\rho}) \cdot (\boldsymbol{\tau} \otimes \boldsymbol{\tau})(\mathbf{v}_s - \mathbf{v}) ds \end{aligned} \tag{14}$$

where $\mathbf{v}_s = \frac{d\boldsymbol{r}_s}{dt}$ is the absolute velocity of the slave point, $\mathbf{v} = \frac{d\boldsymbol{\rho}}{dt}$ is the velocity of the point with CPP on the master surface, κ is the curvature, and $H(*)$ is the Heaviside function.

Equation 14 has three components as mentioned in Ref. [13]. These components include the primary part, the rotation part and the curvature part sequentially. If deformation at the interface is small, the rotation and curvature parts can be ignored.

Additionally, the curvature part vanishes if linear finite elements are used because the curvature of a straight line is zero.

8 Directional Derivative of the Tangential Part

Contact statuses in the tangential direction are distinguished as "sticking" and "sliding". For the sticking status, linearization has the following result:

$$
\begin{aligned}
D_v\left(\delta W_T^c\right) = & -\int_s \frac{\epsilon_T}{\boldsymbol{\rho}_1 \cdot \boldsymbol{\rho}_1} (\delta \boldsymbol{r}_s - \delta \boldsymbol{\rho}) \cdot (\boldsymbol{\rho}_1 \otimes \boldsymbol{\rho}_1)(\mathbf{v}_s - \mathbf{v}) ds \\
& -\int_s \frac{T^{real}}{\left(\boldsymbol{\rho}_1 \cdot \boldsymbol{\rho}_1\right)^2} \left[(\delta \boldsymbol{r}_s - \delta \boldsymbol{\rho}) \cdot (\boldsymbol{\rho}_1 \otimes \boldsymbol{\rho}_1) \frac{\partial \mathbf{v}}{\partial \xi} + \delta \boldsymbol{\rho}_1 (\boldsymbol{\rho}_1 \otimes \boldsymbol{\rho}_1)(\mathbf{v}_s - \mathbf{v}) \right] ds \\
& -\int_s \frac{T^{real} h_{11}}{\left(\boldsymbol{\rho}_1 \cdot \boldsymbol{\rho}_1\right)^2} (\delta \boldsymbol{r}_s - \delta \boldsymbol{\rho})(\boldsymbol{\rho}_1 \otimes \mathbf{n} + \mathbf{n} \otimes \boldsymbol{\rho}_1)(\mathbf{v}_s - \mathbf{v}) ds
\end{aligned}
\tag{15}
$$

Similar to Eq. 14, the second and third lines in Eq. 15 are the rotation and curvature parts respectively, both of which can be ignored if the contact deformation is small. Obviously, the tangent stiffness for the sticking pair is symmetric if only the primary part of Eq. 15 is used.

For the sliding status, we have the directional derivative of δW_T^c as the following:

$$
\begin{aligned}
D_v\left(\delta W_T^c\right) = & -\int_s \frac{\epsilon_N \mu \cdot sign\left(T_1^{trial}\right)}{\sqrt{\boldsymbol{\rho}_1 \cdot \boldsymbol{\rho}_1}} (\delta \boldsymbol{r}_s - \delta \boldsymbol{\rho}) \cdot (\boldsymbol{\rho}_1 \otimes \mathbf{n})(\mathbf{v}_s - \mathbf{v}) ds \\
& -\int_s \frac{\mu |t_N| \cdot sign(T^{real})}{\left(\boldsymbol{\rho}_1 \cdot \boldsymbol{\rho}_1\right)^{3/2}} \left[(\delta \boldsymbol{r}_s - \delta \boldsymbol{\rho}) \cdot (\boldsymbol{\rho}_1 \otimes \boldsymbol{\rho}_1) \frac{\partial \mathbf{v}}{\partial \xi} + \delta \boldsymbol{\rho}_1 (\boldsymbol{\rho}_1 \otimes \boldsymbol{\rho}_1)(\mathbf{v}_s - \mathbf{v}) \right] ds \\
& -\int_s \frac{\mu h_{11} \cdot sign(T^{real})}{\left(\boldsymbol{\rho}_1 \cdot \boldsymbol{\rho}_1\right)^{3/2}} (\delta \boldsymbol{r}_s - \delta \boldsymbol{\rho})(2\boldsymbol{\rho}_1 \cdot \mathbf{n} + \mathbf{n} \otimes \boldsymbol{\rho}_1)(\mathbf{v}_s - \mathbf{v}) ds
\end{aligned}
\tag{16}
$$

Again, the second and third components in Eq. 16 can be omitted when the contact deformation is small.

h_{11} in Eqs. 15 and 16 is the covariant component of the curvature tensor. The primary part of the tangential stiffness for the sliding pair is non-symmetric as indicated in Eq. 16. Consequently, the global tangent stiffness is also non-symmetric.

9 Finite Element Discretization

In the procedure of finite element discretization, original continuous domains are replaced by finite number of small quadrilateral or triangular elements. In contact analysis, three methods are developed in the discrete level: node-to-node,

node-to-surface, and surface-to-surface. Node-to-node approach is the simplest one, but applications are limited to small sliding cases. Node-to-surface or node-to-segment is applicable to contact problem subjected to finite sliding and large deformations. The only limitation is that the slave body has to be approximated by linear finite elements. The surface-to-surface approach is the latest approach which has no limitation of quadratic approximation. This approach is still under development but represents the direction of computational contact analysis. However, this approach is more complicated than the node-to-surface approach in both theory and programming. In this paper, the node-to-segment approach is employed which is suitable for most contact problems in geotechnical engineering.

The discrete version of a contact pair includes one finite element node from the slave body and one line segment from the master body. In 2D cases, independent coordinates in the local coordinate system reduce to one parameter ξ^1. Let $\boldsymbol{x}_1$ and $\boldsymbol{x}_2$ represent position vectors of the start and end nodes of the master segment (Fig. 3), Closest Point Projection (CPP) has a close form solution as follows:

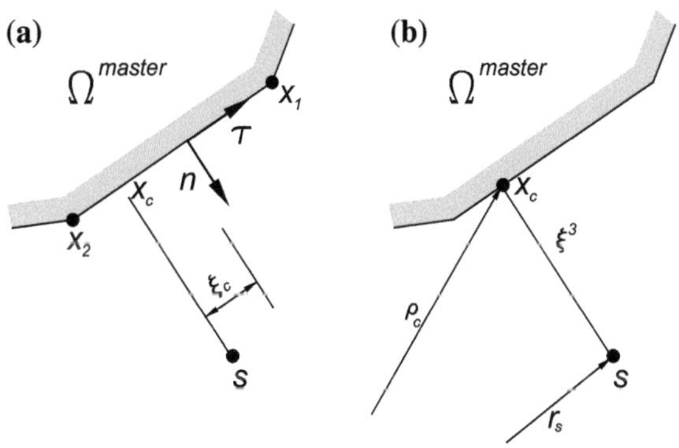

Fig. 3. Depiction of contact pair and location of closest point projection (CPP)

$$\xi_c = \frac{2\boldsymbol{r}_s \cdot (\boldsymbol{x}_2 - \boldsymbol{x}_1) - \boldsymbol{x}_2 \cdot \boldsymbol{x}_2 + \boldsymbol{x}_1 \cdot \boldsymbol{x}_1}{(\boldsymbol{x}_2 - \boldsymbol{x}_1) \cdot (\boldsymbol{x}_2 - \boldsymbol{x}_1)} \tag{17}$$

Points on the master segment can be interpolated by

$$\boldsymbol{\rho}(\xi^1) = \frac{1 - \xi^1}{2}\boldsymbol{x}_2 + \frac{1 + \xi^1}{2}\boldsymbol{x}_1 \tag{18}$$

where $\xi^1 = -1$ at $\boldsymbol{x}_2$ and $\xi^1 = 1$ at $\boldsymbol{x}_1$.

The position vector from the slave point to the master surface is determined by

$$\boldsymbol{r}_s - \boldsymbol{\rho}\left(\xi^1\right) = \begin{bmatrix} -1 & 0 & \frac{1-\xi^1}{2} & 0 & \frac{1+\xi^1}{2} & 0 \\ 0 & -1 & 0 & \frac{1-\xi^1}{2} & 0 & \frac{1+\xi^1}{2} \end{bmatrix} \begin{bmatrix} x_s \\ y_s \\ x_2 \\ y_2 \\ x_1 \\ y_1 \end{bmatrix} = [\mathbf{N}]\{\mathbf{x}\} \tag{19}$$

By means of Eqs. 18 and 19, the tangent stiffness of Eqs. 13–16 can be rewritten to the discrete form. If deformation is small, only the primary parts are reserved. The normal part of the tangent stiffness matrix reads as

$$[\mathbf{K}_N] = \epsilon_N H\left(-\xi^3\right)\left[\mathbf{N}^T\right]\mathbf{n} \otimes \mathbf{n}[\mathbf{N}] \tag{20}$$

Similarly, the tangent stiffness matrix for the sticking pair is revised as

$$\left[\mathbf{K}_T^{stick}\right] = \epsilon_T\left[\mathbf{N}^T\right]\boldsymbol{\tau} \otimes \boldsymbol{\tau}[\mathbf{N}] \tag{21}$$

and the sliding pair has tangent stiffness as

$$\left[\mathbf{K}_T^{slide}\right] = \mu\epsilon_N \frac{\Delta\xi^1}{\left|\Delta\xi^1\right|}\left[\mathbf{N}^T\right]\boldsymbol{\tau} \otimes \mathbf{n}[\mathbf{N}] \tag{22}$$

Assembling of these contact stiffness $[\mathbf{K}_N]$, $\left[\mathbf{K}_T^{stick}\right]$ and $\left[\mathbf{K}_T^{slide}\right]$ into the global system tangent matrix follows standard procedures.

10 Contact Smoothing

Curved boundaries are not smooth after being discretized by linear finite elements. If structures have a rectangular shape, the corners are also not smooth. These non-smooth profiles have following issues:

a. Normal and tangential units jump from one segment to its adjacent segment.
b. Normal and tangential units have no unique value at the interaction node.
c. Unrealistically large penetration may occur at corners of the rectangle.

This non-smoothing transition of local unit vectors often triggers convergence issues. A typical example is occurrence of “chattering” when the slave node slides back and forth around an intersection on the master surface. In this paper, a Hermite interpolation method is used to smooth the contact interface as depicted in Fig. 4. Segments x_2x_3 and x_3x_4 are adjacent FE edges on the master body. Point x_a locates at αL_{23}, where $0 \leq \alpha \leq 0.5$ and L_{23} is the length of segment x_2x_3. Tangents at ends of this interpolated curve x_ax_b are defined as

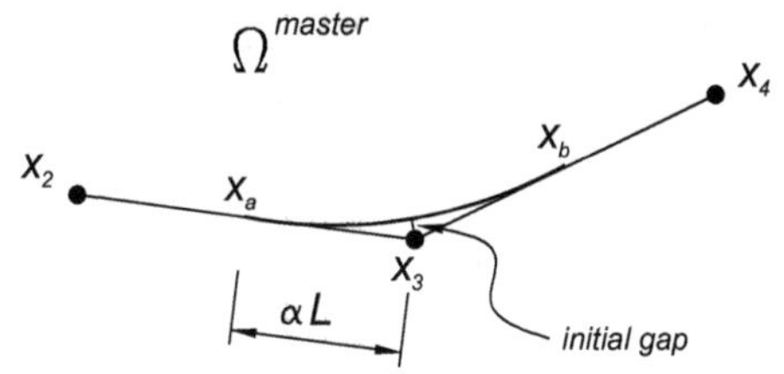

Fig. 4. Smoothing C^0 continuous boundary by Hermite interpolation

$$\frac{d\boldsymbol{x}_a}{d\xi} = \boldsymbol{x}_3 - \boldsymbol{x}_a, \quad \frac{d\boldsymbol{x}_b}{d\xi} = \boldsymbol{x} - \boldsymbol{x} \tag{23}$$

The Hermite curve is interpolated by

$$\begin{aligned} \boldsymbol{x} = & \frac{1}{4}(\xi^3 - 3\xi + 2)\boldsymbol{x}_a + \frac{1}{4}(-\xi^3 + 3\xi + 2)\boldsymbol{x}_b + \frac{1}{4}(\xi^3 - \xi^2 - \xi + 1)\frac{d\boldsymbol{x}_a}{d\xi} \\ & + \frac{1}{4}(\xi^3 + \xi^2 - \xi + 1)\frac{d\boldsymbol{x}_b}{d\xi} \end{aligned} \tag{24}$$

where ξ is the local coordinate of the interpolated curve $x_a x_b$.

When Hermite smoothing or other high order smoothing algorithm is used, CPP procedure should use the Newton method to find ξ_c.

11 Numerical Examples

Two examples in geotechnical engineering are presented in this section. Examples are analyzed by the FEM software package En*FEM* which was developed following implementations in this paper. Quadrilateral linear four-node elements are used for both examples.

12 Embedded Culvert Subject to Gravity Load and Unsymmetrical Ground Surcharge

The initial configuration of this example is illustrated in Fig. 5. The culvert is made up of concrete with Young's modulus 2.8×10^7 kPa. This culvert has a 180 cm outer diameter and a 144 cm inner diameter. The soil is modelled using the Mohr-Coulomb elasto-plastic material model. The coefficient of friction between the concrete and soil is 0.6. Applied loads include the gravity load and a surcharge load of 35 kPa/per unit length. Since the ground surcharge is not symmetrical, full model of culvert is used in simulation.

Two stages are considered in this analysis: stage one, with gravity load applied, and stage two, with the additional ground surcharge applied. The normal contact pressure and friction stress at the end of each stage are plotted in Figs. 6 and 7 respectively.

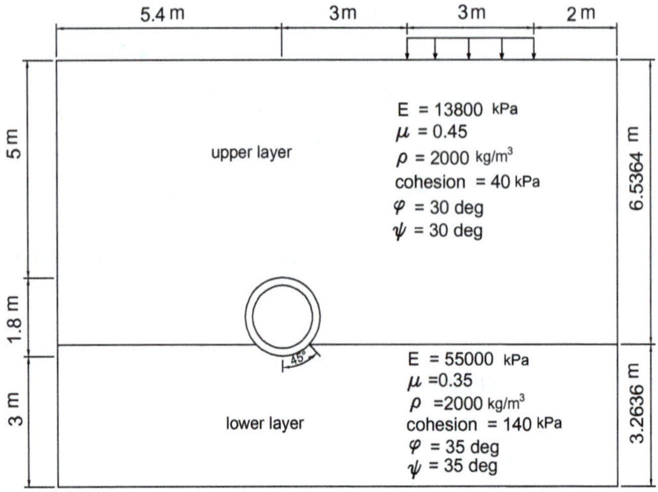

Fig. 5. Embedded culvert subject to gravity and unsymmetrical surcharge

Note that central angles start from the culvert top and rotate in the clockwise direction. Dashed orange lines are results at the end of stage 1. After the surcharge is applied, distribution of contact pressure is no more symmetric. The minimum normal pressure locates around the right interface between soft and stiff soil layers. Moment and thrust forces at the end of each stage are plotted in Figs. 8 and 9 respectively. Distributions of moment and thrust forces for each stage follow the same oscillation rhythm with shift of peaks due to the unsymmetrical total loads.

Relative sliding between soil and the culvert is very small, so that all contact pairs are sticking. No gaps are developed in this simulation.

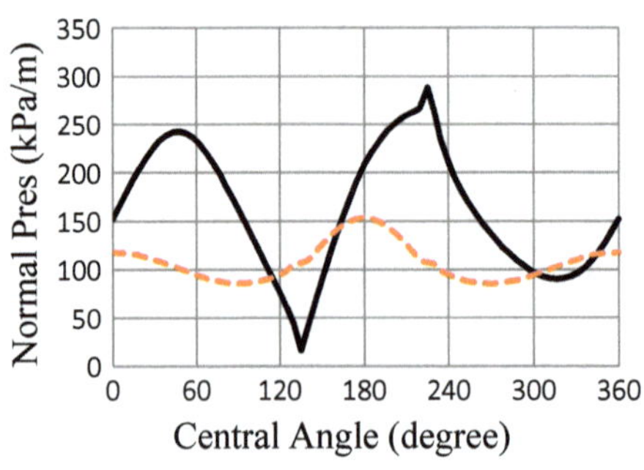

Fig. 6. Contact pressure

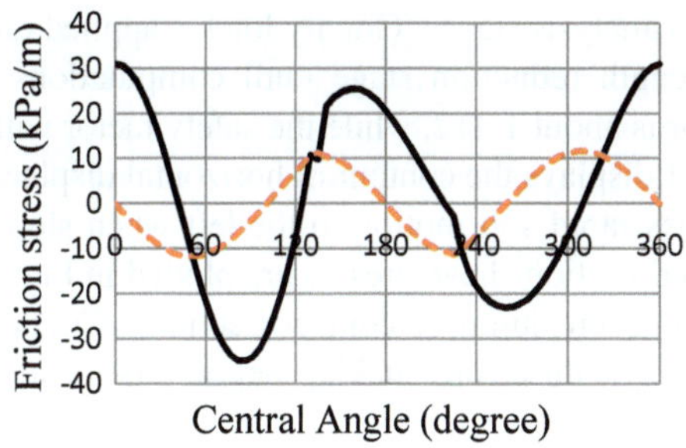

Fig. 7. Friction stress

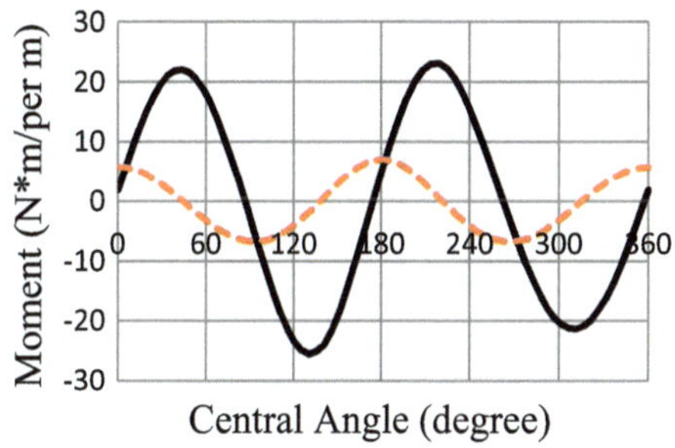

Fig. 8. Moment of culvert

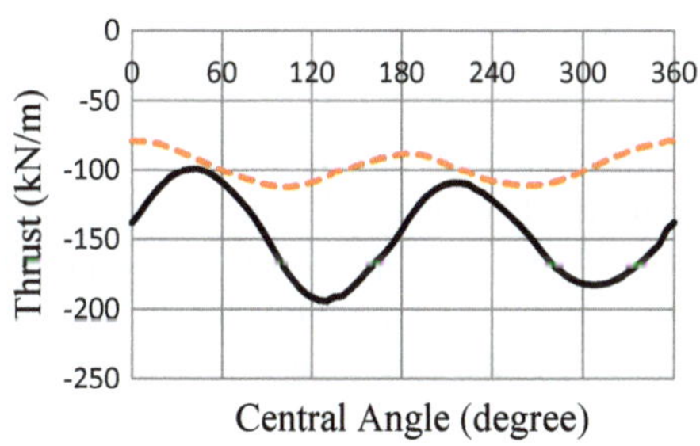

Fig. 9. Thrust of culvert

13 Stability Analysis of Slope Strengthened by a Stabilizing Pile (2D)

The second example is a 2D stability analysis of slope strengthened by a stabilizing pile. The strength reduction method is used to estimate the factor of safety against sliding. The stabilizing pile has 20 cm width and 15.6 m of length. Embedded length into the lower soil layer is about 5 m. Material properties and model geometry are shown in Fig. 10. The Mohr-Coulomb elasto-plastic material model is used for the soil.

This model has two analysis stages. Gravity load is applied during the first analysis stage followed by strength reduction stage until computations no longer converge. Calculated safety factor is about 1.412, while the safety factor without a stabilizing pile is about 1.31. Figure 11 displays the contour of horizontal displacements at failure. The deep blue zone indicates rapid soil motion to the left when slope failure occurs.

Contact pressure and friction shear stresses are plotted in Fig. 12. Approximately at depth 10 m, there is a severe jump of contact tractions. Contact statuses above this location are sliding while statuses are sticking below this location.

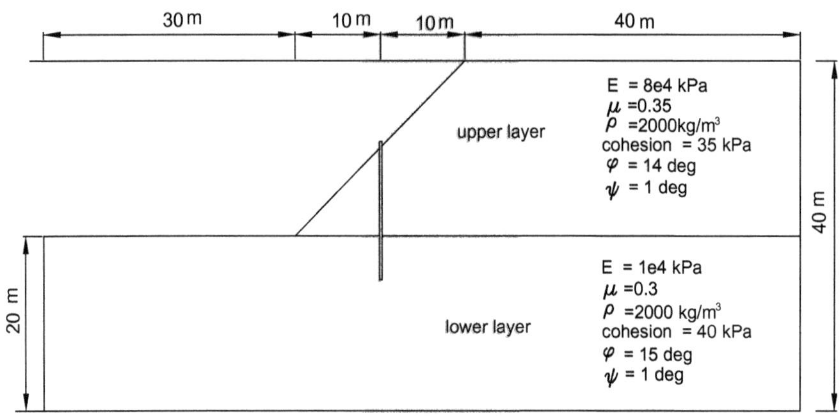

Fig. 10. Stability analysis of slope strengthened by a stabilizing pile

14 Conclusions

This paper presents implementation of computational contact analysis. Formulations in covariant form show unique advantages of being concise and easy to understand. Although non-penetration is satisfied approximately, the penalty approach provides convenient implementation and robust performance in contact problems involving finite sliding. Experience has found that local smoothing is essential if linear finite elements are used because the node-to-segment approach is sensitive to non-smooth interfaces. Two examples of 2D soil-structure interaction problems are presented.

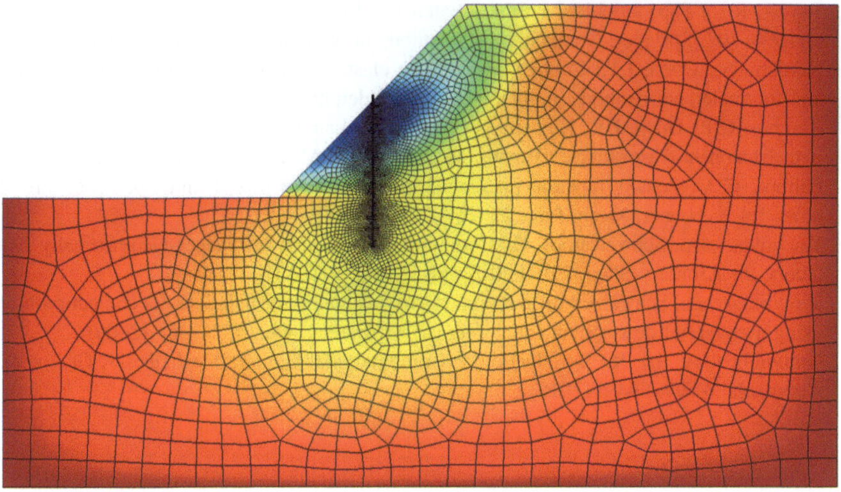

Fig. 11. Contour of horizontal displacements at slope failure

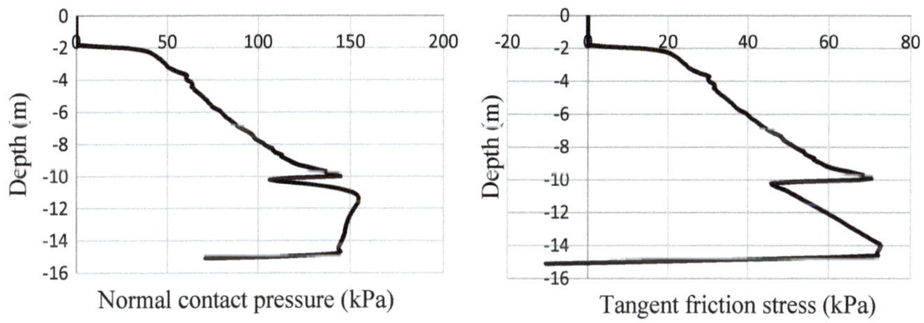

Fig. 12. Normal pressure and friction stress on right side of pile

These examples were solved using a FEM software package that was developed using the recommendations of this paper.

References

1. Goodman, R.E., Taylor, R.L., Brekke, T.L.: A model for the mechanics of jointed rock. J. Soil Mech. Found. Div. ASCE, SM3 **94**, 637–659 (1968)
2. Desai, C.S., Zaman, M.M., Lightner, J.G., Siriwardane, H.J.: Thin-layer elements for interfaces and joints. Int. J. Numer. Anal. Meth. Geomech. **8**, 19–43 (1984)
3. Francavilla, A., Zienkiwicz, O.C.: A note on numerical computation of elastic contact problems. Int. J. Numer. Meth. Eng. **9**, 913–924 (1975)
4. Hughes, T.J.R., Taylor, R.L., Sackman, J.L., Curnier, A., Kanoknukulchai, W.: A finite element method for a class of contact-impact problems. Comput. Methods Appl. Mech. Eng. **8**, 249–276 (1976)

5. Taylor, R.L., Papadopoulos, P.: On a patch test for contact problems in two dimensions, pp. 690–702. Springer, Computational Methods in Nonlinear Mechanics (1991)
6. Kikuchi, N., Oden, J.T.: Contact problems in elasticity: a study of variational inequalities and finite element methods, p. 495. SIAM, Philadelphia (1988)
7. Campos, L.T., Oden, J.T., Kikuchi, N.: A numerical analysis of a class of contact problems with friction in elastostatics. Comput. Methods Appl. Mech. Eng. **34**, 821–845 (1982)
8. Felippa, C.A.: Penalty-function interactive procedures for mixed finite elements formulations. Int. J. Numer. Mech. Eng. **22**, 267–279 (1986)
9. Konyukhov, A., Schweizerhof, K.: Computational Contact Mechanics-Geometrically Exact theory for Arbitrary Shaped Bodies. Springer, 446 p (2013)
10. Konyukhov, A., Schweizerhof, K.: A special focus on 2D formulations for contact problems using a covariant description. Int. J. Numer. Meth. Eng. **66**, 1432–1465 (2006)
11. Konyukhov, A., Izi, R.: Introduction to Computational Contact Mechanics-A Geometrical Approach. Wiley, 304 p (2015)

Performance of Geocell Reinforced Embankment over Soft Soil Deposit

Lalima Banerjee[1,2], Sowmiya Chawla[1(✉)], and Gupinath Bhandari[3]

[1] Department of Civil Engineering, Indian Institute of Technology (Indian School of Mines), Dhanbad, Jharkhand, India
lalima.banerjee@gmail.com, sowmiya_iitd@hotmail.com
[2] School of Civil Engineering, KIIT University, Bhubaneswar, India
[3] Department of Civil Engineering, Jadavpur University, Kolkata West Bengal, India
g.bhandari@civil.jdvu.ac.in

Abstract. This paper studies the performance improvement due to the presence of geocell layer at the base of the embankment over soft clay deposit. The experimental study was carried out by conducting tests on model embankments overlying soft clay with reinforcement by geocell mattress at the interface between the earthen embankment and soft clay foundation bed, by varying its parameters like aspect ratio (pocket size), and type of infill material. Experiments were also carried out for unreinforced foundation soil beneath the earthen embankment, to show comparative performance and ground improvement due to soil reinforcement using geocell mattress. Test results indicated a considerable percentage reduction in settlements, lateral deformations and percentage increase in ultimate capacity of geocell reinforced embankments, over unreinforced embankment. The test results were also validated using finite element analyses of geocell reinforced foundation over unreinforced foundation.

1 Introduction

Construction of embankment or other geotechnical structures over soft soil has always been a challenge for engineers or contractors. The difficulty arises mainly due to large settlements and lateral deformations that take place due to the extremely low shear strength of the soft foundation soil. Methods like soil excavation and replacement or piling are not practical or economical in many cases. In this aspect, the use of geocells has been a major contribution as a ground improvement technique. Geocell reinforcement is generally provided at the foundation level, however it can also be effectively provided as a slope reinforcement technique owing to its three-dimensional structure. Geocells are three-dimensional, honey-combed like, polymeric structures which are interconnected at their joints. Due to its 3-D structure, it can contain the soil at the base of the embankment, which helps to distribute the load over a wider area by providing a stiff and rigid base to the embankment and thus, improving the

M.-C. Weng et al. (eds.), *Current Geotechnical Engineering Aspects of Civil Infrastructures*, Sustainable Civil Infrastructures,
https://doi.org/10.1007/978-3-319-95750-0_6

performance of embankments over soft soil in terms of increased bearing capacity and reduced settlements and lateral deformations. From literature, it has been also seen that it is economical as well as much less time consuming than piling or other conventional methods.

Use of Geocells has been established to be an effective reinforcement technique by experimental studies in the laboratory (e.g. Rea and Mitchell 1978; Mhaiskar and Mandal 1994; Emersleben and Meyer 2008), by numerical modeling (Mehdipour et al. 2013; Leshchinsky and Ling 2013) and also by conducting field studies of geocell applications (Paul 1988; Sitharam and Hegde 2013). Mandal and Gupta (1994), Dash et al. (2001a, b, 2004), Tafreshi and Dawson (2012) reported the beneficial effects of using geocells on strip footing by conducting laboratory model tests. Krishnaswamy et al. (2000) observed that geocell used at the embankment base overlying soft clay could greatly reduce the deformations for various applied surcharge pressures. Dash et al. (2003a, b) conducted laboratory model tests on circular footings using geocells to investigate the improvement in performance of circular footings. Latha et al. (2006), Zhang et al. (2010) studied the advantages of geocell application at the base of embankment over soft foundation soil by conducting laboratory model embankment tests. The confinement effect provided by the geocells on sand samples were studied by triaxial compression tests and found that a large amount of apparent cohesion is developed in the reinforced samples (e.g. Rajagopal et al. 1999; Chen et al. 2013) which showed that the shear strength of soils increased using geocells. Chen and Chiu (2008) conducted laboratory model tests on geocell retaining walls and found considerable reduction in settlement and lateral displacement. Pokharel et al. (2009), Pokharel et al. (2010) carried out laboratory tests to study the improvements in bearing capacity of base course using geocell. Hegde and Sitharam (2017) conducted laboratory studies as well as numerical simulations using three types of cellular reinforcement, namely geogrid cells, commercial geocells and bamboo cells and obtained the best performance using bamboo cells owing to its higher stiffness.

Robertson and Gilchrist (1987) reported the selection of geocell mattress as the most cost effective method for constructing a 4 m high embankment over soft foundation soil. The selection of geocell option was made after economic analysis of excavation and replacement. Other options were not considered because they were either impractical or would take too long to construct. Paul (1988) examined four alternatives for embankment construction over areas in Scotland and found geocell option to be the most economical and rapid. Mehdipour et al. (2013) studied the behavior of geocell reinforced slopes and found that the lateral deformation and shear strain of slope considerably decreased. Leshchinsky and Ling (2013) studied the confinement provided by geocells to railway ballast through numerical modeling and showed that geocell was effective in reducing the stresses on the subgrade by distributing it more uniformly.

The purpose of the present research is to show the improvement in performance of geocell reinforced soft foundation soil of embankment with respect to increase in bearing capacity and decrease in settlements and lateral deformations of embankment and heaving of soft clay foundation under applied loads and validate the results using finite element analyses.

2 Model Tests

2.1 Materials Used

Soft foundation bed was prepared using locally available clayey and organic soil in the ratio 1:3 by weight. The specific gravity of the soft soil was found to be 1.56 by water pycnometer method. The undrained shear strength of soft foundation bed was found by unconfined compression tests. Locally available clayey soil was used as embankment soil and also as infill material in geocells. The effective shear strength parameters of embankment soil were found by consolidated drained triaxial tests. The properties of the soft foundation soil and embankment soil used in the present investigation are given in Table 1. In few tests, dry river sand was also used as infill soil. The properties of the sand used as infill material are given in Table 2. The geocell was constructed using biaxial geogrid. The ultimate tensile strength of the biaxial geogrid was 20 kN/m at failure strain 25%. The properties of geogrid used for geocell construction are given in Table 3. The geocells were constructed having diamond pattern for all the tests.

Table 1. Properties of soft foundation soil and embankment soil used in the present investigation

Property	Soft foundation soil	Embankment soil
L.L. (%)	52	53.9
P.L. (%)	25	22.64
Optimum moisture content (%)	24.91	17.2
Max dry density (kg/m^3)	1320	1540
Mean specific gravity (G)	1.56	2.39
Placement moisture content (%)	40	17.2
Placement density (kg/m^3)	1670	1540
Cohesive strength (kPa)	20	90
Angle of internal friction (°)	–	7

Table 2. Properties of sand used as infill material

Property	Infill sand
Minimum density, ρ_{min} (kg/m^3)	1370
Maximum density, ρ_{max} (kg/m^3)	1700
Uniformity coefficient (Cu)	1.55
Coefficient of curvature (Cc)	0.903
Mean specific gravity (G)	2.69
Porosity (η) (%)	0.2
Maximum void ratio (e_{max})	0.992
Minimum void ratio (e_{min})	0.605

Table 3. Properties of geogrid used for geocell construction

Property	Geogrid
Ultimate tensile strength (kN/m)	20
Failure strain (%)	25
Thickness (mm)	2
Aperture opening shape	diamond

2.2 Test Set-up

A steel tank having dimensions 1150 mm × 550 mm × 975 mm was fabricated for conducting model tests on embankment overlying soft clay. The test set-up was prepared in a two-side Perspex fitted steel tank so that the deformation behavior of the embankment could be clearly observed. Geocell mattress was placed above soft clay bed within the tank and the embankment was constructed above the geocell layer. The test bed was instrumented using proving ring and dial gauges. The load was applied to the embankment using a hydraulic jack supported against a reaction frame. The load was measured using a pre-calibrated proving ring of capacity 10-tonne. The vertical displacements and lateral deformations were measured using dial gauges of sensitivity 0.01 mm and maximum deflection 50 mm. Figure 1 represents the test set-up of the model embankment tests used in the present investigation.

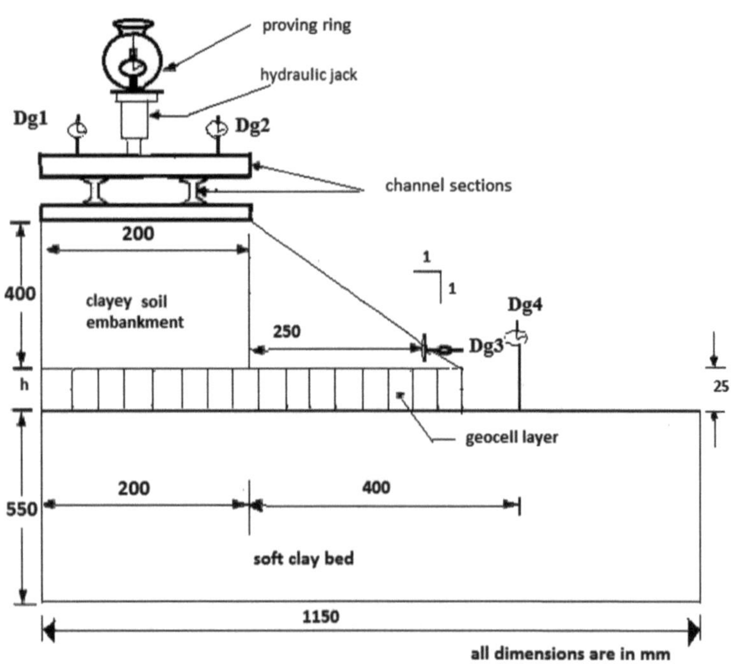

Fig. 1. Test set-up of model embankment tests

2.3 Test Bed Preparation

2.3.1 Preparation of Soft Clay Bed at Foundation

The soft foundation bed was prepared by using clayey and organic soil in the ratio 1:3 by weight. The soil was mixed with predetermined amount of water to maintain same placement moisture content throughout and uniform moisture distribution was obtained. The foundation was prepared by layer-wise proper compaction of the soil to the desired height. The amount of soil needed, water content of soil, height of fall and number of blows of the compacting equipment required to achieve the desired density was determined prior to the model tests through a series of trials. The water content and compaction were carefully controlled to achieve a fairly uniform test condition throughout the entire test program. Each layer was uniformly compacted in order to achieve uniform density in all the tests. Undisturbed samples were taken from each layer to determine the in situ unit weight and moisture content of the foundation soil so that uniformity could be maintained for each test.

2.3.2 Fabrication and Laying of Geocell Mattress

After leveling the soft foundation bed, a layer of geocell mattress was placed at the interface of the soft foundation and embankment which was truncated at the embankment toe. The geocell mattress was prepared by cutting the geogrid to required length and height from full rolls. The geogrids were bound with wires to achieve the desired diamond shape. After the fabrication of geocell layer, the geocell pockets were filled with sand/clay by dropping it from a certain height. The height of fall to achieve the desired relative density was determined beforehand through trials with different heights of fall. Samples were collected in small aluminium cans of known volume to monitor the relative density achieved. The cans were placed at different locations in the test tank and the difference in densities was measured and found to be less than 1%.

2.3.3 Preparation of Clay Embankment

The embankment having a slope 1:1 was prepared by compacting the soil in four equal layers till the desired height was attained. The embankment was compacted at optimum moisture content to achieve the maximum dry density. The amount of soil, number of blows and height of fall were determined prior to the model tests by a series of trial tests. Each layer was compacted using calculated number of blows to achieve uniform density in all the tests. Undisturbed samples were taken during each loading and unloading stage to carefully monitor the water content and compaction to maintain uniformity during each test.

2.4 Test Procedure

The experimental program for the model tests conducted is given in Table 4. Surcharge pressure was applied on the embankment using a hydraulic jack. A 10-tonne proving ring was used to measure the surcharge load. Until the deformation under a particular load increment had stabilized, it was maintained constant. Uniform surcharge pressure distribution was achieved using an arrangement of three steel I-sections between two rigid steel plates running for full width of the tank on the embankment crest. Four dial gauges were placed at different locations to measure vertical and horizontal deformations (Fig. 2 a, b). Two dial gauges (Dg1 and Dg2) measured vertical settlements (Fig. 2a); Dg3 measured lateral deformations near the embankment toe and Dg4 measured the heaving of the soft foundation under applied load (Fig. 2b). For each test, the pressure versus deformation readings were taken till the ultimate capacity was reached. The laboratory model tests continued for approximately 30 min. After each test, the foundation and embankment soil were completely unloaded and again freshly prepared for the next test. The moisture content and density were maintained uniformly by taking undisturbed samples for each loading and unloading stage. The geocell mattress was also carefully removed after completion of each test. The trial test results on reinforced and unreinforced ensured uniformity of test conditions.

Table 4. Experimental program for the model tests conducted

Type of material	Pattern	Thickness (mm)	Type of infill material	Aperture size (mm)	Aspect ratio	No. of tests
Unreinforced	–	–	–	–	–	1
Geocell	Diamond	25	Clay	75 × 75	0.33	1
				100 × 100	0.25	1
			Sand	75 × 75	0.33	1
				100 × 100	0.25	1

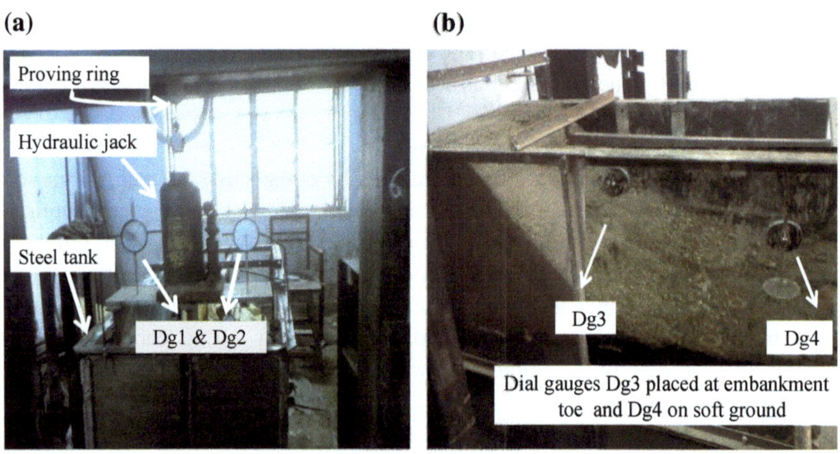

Fig. 2. Dial gauges placed at different location in the model tests: **a** Dg1 and Dg2 for vertical settlements; **b** Dg3 at embankment toe and Dg4 on soft foundation for heaving measurement

2.5 Model Test Results

Since the model tests were completed in approximately 30 min, the short term behavior of the embankment under applied load was considered in this study. Figure 3 shows the pressure-settlement behavior of unreinforced and geocell reinforced embankments for different aspect ratios and infill materials. From the model test results, it was observed that for unreinforced case, the pressure-settlement curve slope became almost vertical beyond a settlement of 7.5% of embankment top width, indicating failure. It can be seen that the clay-filled geocells (h/D = 0.25) gives an ultimate capacity of 218.72 kPa as compared to the unreinforced embankment (82 kPa) which means that 166.67% increase in ultimate capacity was achieved using clay-filled geocells. Also at the ultimate capacity of unreinforced embankment, clay-filled geocells (h/D = 0.25) gave 87.9% reduction in vertical settlement. For sand-filled geocells having same aspect ratio (h/D = 0.25), increase in ultimate capacity was 6.25% over clay-filled geocells and further reduction in vertical settlement was 11.27%.

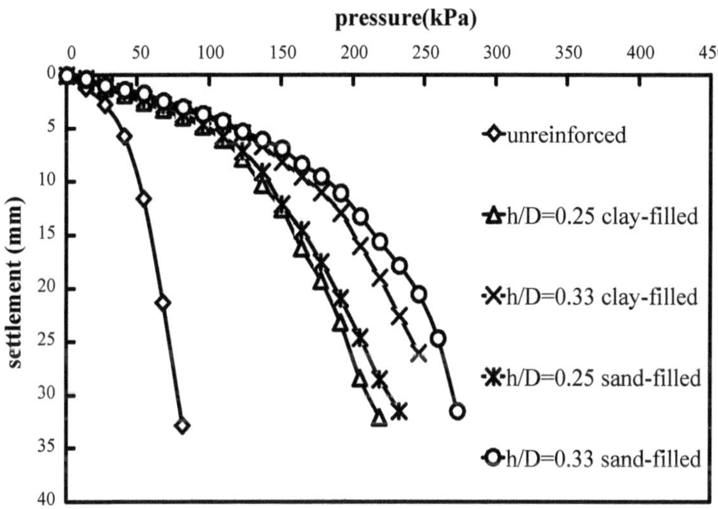

Fig. 3. Pressure-settlement behavior of model embankments for different aspect ratios and infill materials

For clay-filled geocells having higher aspect ratio 0.33, increase in ultimate capacity over 0.25 sand-filled geocells was 5.88%. The reduction in settlement for clay-filled geocells (h/D = 0.33) was 28.25% over sand-filled geocells (h/D = 0.25) at the ultimate capacity of 0.25 sand-filled geocells. For same aspect ratio (h/D = 0.33), sand-filled geocells gave further increase in ultimate capacity of 11.11% over clay-filled geocells. It was observed that the pressure-settlement curves were much stiffer than unreinforced case, indicating that performance substantially improved using geocells.

Figure 4 shows the lateral deformation at the toe of the embankment for different values of applied pressure. It is observed that lateral deformation is considerably reduced using geocell as reinforcement irrespective of the infill material. This happens due to the frictional resistance offered by the composite geocell-infill system. However for the same aspect ratio, sand-filled geocells gave better results as compared to clay-filled geocells. A higher aspect ratio increased the performance of geocell irrespective of the infill material.

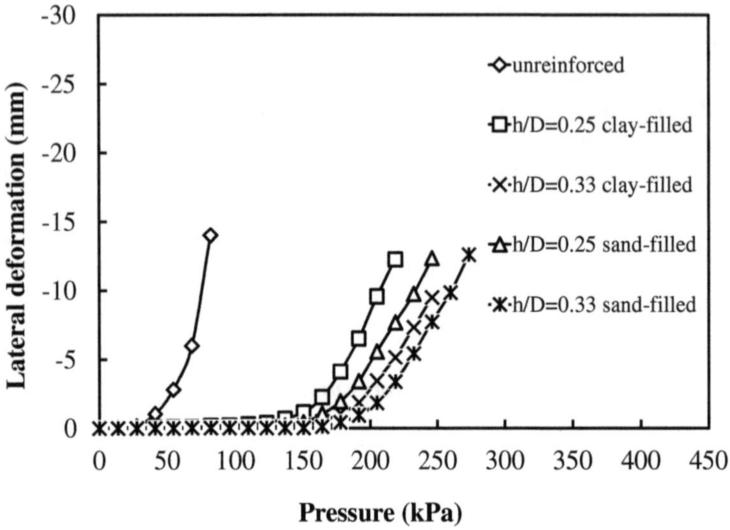

Fig. 4. Pressure-lateral deformation behavior of model embankments for different aspect ratios and infill materials at embankment toe

Figure 5 shows the heaving of soft foundation bed under applied pressure during the model tests. The heaving of the soil also follows the same pattern i.e. it reduces using geocells of higher aspect ratio and sand as infill material.

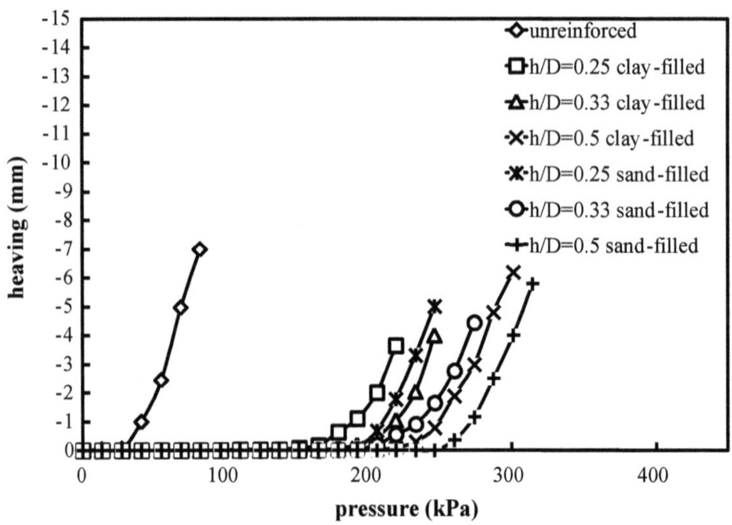

Fig. 5. Pressure-heaving behavior of soft soil bed for different aspect ratios and infill materials

Thus, the highest load carrying capacity was achieved by using geocell having 75 mm pocket size (h/D = 0.33) i.e., smaller pocket size. The different aspect ratios were achieved by varying only the pocket size as height of geocell layer was kept constant in all tests (25 mm). The decrease in pocket size results in increase in confinement of cells per unit volume and increase in rigidity of geocell mattress which results in the performance improvement. Sand as infill material can be compacted with better stiffness and rigidity than clay. However, in absence of good quality sand near construction site, locally available clay can also be used as infill material.

3 Fem Modelling

Midas GTS Software (Midas 2013) was used to perform three-dimensional finite element analyses of model embankments. Midas GTS is a commercially available, fully integrated 2D and 3D finite element software. In this investigation, straight analyses have been used to validate the test results which implies that while the soft foundation was modeled using elasto-plastic, Mohr-Coulomb, undrained relationship, the embankment soil and infill material were modeled using elasto-plastic, Mohr-Coulomb, drained relationship.

3.1 Geometry and Mesh Generation

For validating the model test results, a three dimensional finite element model of exactly the same geometry as laboratory model embankment was prepared using Midas GTS. The linear elastic model was used to simulate the loading plate. The modulus of elasticity for the loading plate and Poisson's ratio were taken from Sowmiya (2013). The geometry of the geocell layer of individual pocket size 100 mm × 100 mm was prepared with the required diamond shape over the surface of the foundation soil. It was then extruded as a solid element of required height. The geocell was modelled as a linearly elastic material. The Young's modulus of geocell used was 120 MPa from a range of values of geocell stiffness used by Leschinsky and Ling (2013). The Poisson's ratio for the geocell was assumed to be 0.2. The undrained Mohr-Coulomb model was used to model the soft foundation while the drained Mohr-Coulomb model was used for the infill soil and embankment. The undrained shear strength parameters for the foundation soil and the effective shear strength parameters for the embankment soil used were the same as found experimentally as already given in Table 1. The moduli of elasticity and Poisson's ratio for the embankment soil, infill sand and soft foundation soil were taken from Bowles (1996). The effective angle of internal friction for infill sand was found by direct shear tests. The cohesion value was assumed to be 0.1 kPa for ease of computation. After geometry modeling, the mesh was generated using auto-mesh generation option having triangular elements. A desired element size with adaptive seeding was used for this purpose. The model components were separated using 'Boolean Cut' option before auto-mesh generation. Figure 6a–f shows all the model embankment components used for the three-dimensional finite element model. The parameters assigned to each mesh set of individual embankment components for straight analyses are given in Table 5.

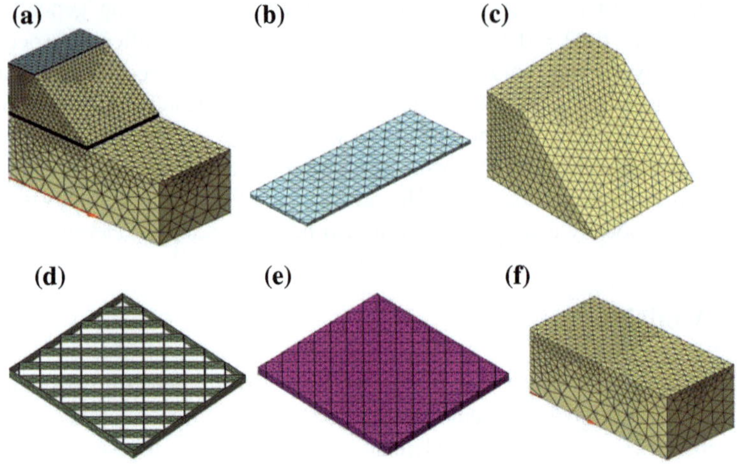

Fig. 6. Components of model embankment: **a** model embankment with all layers; **b** loading plate; **c** embankment soil layer; **d** geocell; **e** infill soil; **f** soft foundation

Table 5. Parameters for straight analyses of FEM model

Model components	Straight analysis model parameters used in FEM	E (MPa)	μ	(c_u) (kPa)	ϕ′ (°)
Loading plate	Elastic	200,000	0.27	–	–
Embankment soil	Mohr coulomb (drained)	6	0.3	90	7
Sand	Mohr coulomb (drained)	30	0.3	0.1	30
Soft foundation clay	Mohr coulomb (undrained)	2	0.4	20	–
Geocell	Elastic	120	0.2	–	–

3.2 Boundary Conditions

Roller supports were assigned to the vertical faces of embankment model meaning that the side boundaries of the model in the tank were restrained only in the horizontal direction while vertical displacements were allowed to occur, but no boundary conditions were applied to the sloping surfaces of the embankment layers. The bottom face of the subgrade was considered fixed. The default values of most of the computational control parameters were kept same as given in Midas GTS except displacement norm, which was changed to 0.01.

3.3 Validation of Test Results Using FEM

The results of the model embankment tests were compared with that of FEM results. A total of two tests using straight analyses were simulated for geocells having aspect ratio 0.25 with sand and clay as infill material. The tests were completed in 30 min. Figure 7 shows the comparison of FEM results with the measured results of laboratory model tests for aspect ratio of 0.25, having sand and clay as infill material. The FEM analysis gives fairly similar behavior as compared to the model test results. Figures 8a, b show the distribution of total displacements and vertical stresses for different model embankment components for sand-filled geocell (h/D = 0.25). The displacement contours portray the efficiency of geocell in reducing vertical displacements while the vertical stress contours show that the incorporation of geocell leads to minimizing the stresses gradually in the subsequent layers leading to a shift from punching to local shear failure. The presence of a geocell layer shifts the potential failure planes further downwards leading to increase in ultimate capacity. From the FEM results, it is clear that the geocells having sand as infill material gives better results than clay as infill material. Also higher aspect ratios give better performance in terms of ultimate capacity and reduction in settlements, lateral deformations and heaving of soft foundation bed.

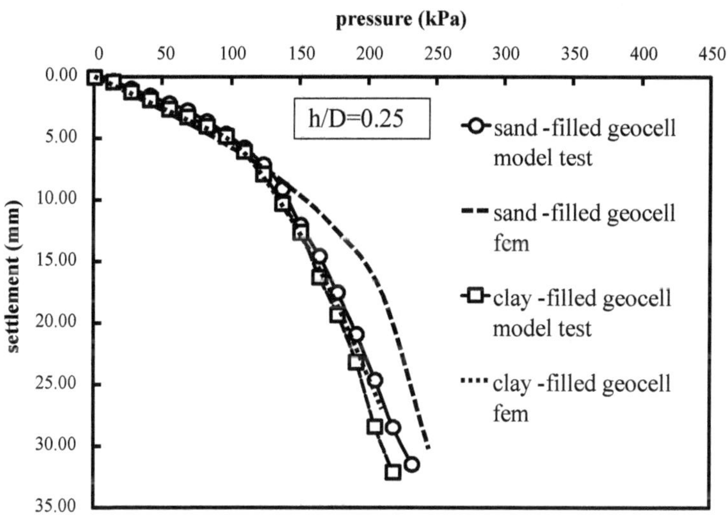

Fig. 7. Comparison of pressure-settlement behavior of laboratory model embankment tests with FEM for geocells with 0.25 aspect ratios and sand and clay as infill materials

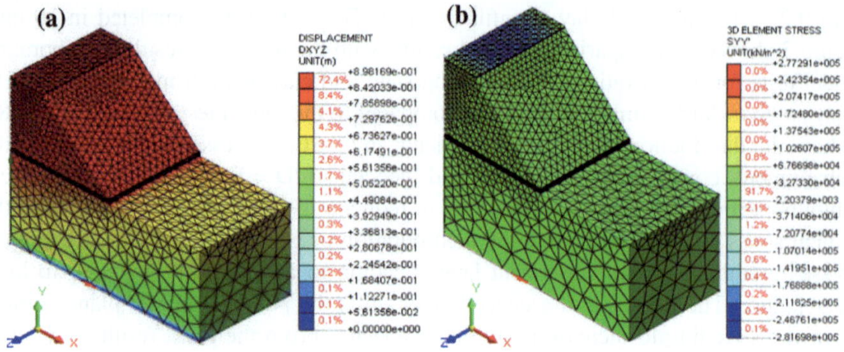

Fig. 8. Distribution of **a** total displacement and **b** vertical stress for sand-filled geocell having aspect ratio 0.25

4 Conclusions

Laboratory model tests were conducted on model embankments and FEM modeling using straight analyses were performed. From the present study, the following conclusions can be drawn:

1. The inclusion of geocell mattress at the base of embankment can increase its ultimate capacity by about 3 times to that of unreinforced embankment.
2. The improvement in performance increased when geocells with higher aspect ratio (smaller pocket size) was used.
3. For the same aspect ratio, geocells having sand as infill material gave better results than clay, both in terms of increase in ultimate capacity and decrease in settlements, lateral deformations and heaving of soft foundation.
4. The settlements, lateral deformations and heaving of soft foundation were greatly reduced using geocell mattress.

References

Bowles, J.E.: Foundation Analysis and Design, 5th edn. McGraw-Hill International Edition (1996)

Chen, R.H., Chiu, Y.M.: Model tests of geocell retaining structures. Geotext. Geomembr. Elsevier (2008). https://doi.org/10.1016/j.geotexmem.2007.03.001

Chen, R.H., et al.: Confinement effect of geocells on sand samples under triaxial compression. Geotext. Geomembr. Elsevier (2013). https://doi.org/10.1016/j.geotexmem.2013.01.004

Dash, S.K., et al.: Bearing capacity of strip footings supported on geocell-reinforced sand. Geotext. Geomembr. Elsevier (2001a). https://doi.org/10.1016/S0266-1144(01)00006-1

Dash, S.K., et al.: Strip footing on geocells reinforced sand beds with additional planar reinforcement. Geotext Geomembr Elsevier (2001b). https://doi.org/10.1016/S0266-1144(01)00022-X

Dash, S.K., et al.: Model studies on circular footing supported on geocell reinforced sand underlain by soft clay. Geotext Geomembr, Elsevier (2003a). https://doi.org/10.1016/S0266-1144(03)00017-7

Dash, S.K., et al.: Behaviour of geocell- reinforced sand beds under circular footing. Ground Improv. **7**(3), 111–115 (2003b). https://doi.org/10.1680/grim.2003.7.3.111

Dash, S.K., et al.: Performance of different geosynthetic reinforcement materials in sand foundations. Geosynthetics Int. **11**(1), 35–42 (2004). https://doi.org/10.1680/gein.2004.11.1.35

Emersleben, A., Meyer, N.: Bearing capacity improvement of asphalt paved road constructions due to the use of geocells—falling weight deflectometer and vertical stress measuresments. Geosynthetics Environ. Eng. (2008). https://doi.org/10.1007/978-3-540-69313-0_136

Hegde, A., Sitharam, T.G.: Experiment and 3D-numerical studies on soft clay bed reinforced with different types of cellular confinement systems. Transp. Geotech. Elsevier (2017). https://doi.org/10.1016/j.trgeo.2017.01.001

Krishnaswamy, N.R. et al.: Model studies on geocell supported embankments constructed over a soft clay foundation. Geotech. Test. J. ASTM (2000). https://doi.org/10.1520/gtj11122j

Leshchinsky, B., Ling, H.I.: Numerical modelling of behavior of railway ballasted structure with geocell confinement. Geotext. Geomembr. Elsevier (2013). https://doi.org/10.1016/j.geotexmem.2012.10.006

Latha, M., et al.: Experimental and theoretical investigations on geocell-supported embankments. Int. J. Geomech. ASCE (2006). https://doi.org/10.1061/(ASCE)1532-3641(2006)6:1(30)

Mandal, J.N., Gupta, P.: Stability of geocell-reinforced soil. Constr. Build. Mater. **8**(1), 55–62 (1994). https://doi.org/10.1016/0950-0618(94)90009-4

Mehdipour, et al.: Numerical study on stability analysis of geocell reinforced slopes by considering the bending effect. Geotext. Geomembr. Elsevier (2013). https://doi.org/10.1016/j.geotexmem.2013.01.001

Mhaiskar, S.Y., Mandal, J.N.: Soft clay sub grade stabilisation using geocells. In: Mandal, J.N. (ed.) Geosynthetic World, pp. 139–148. Wiley Eastern Ltd. (1994)

Midas, G.T.S.: Geotechnical and Tunnel Analysis System reference manual for Modelling. Midas Corporation, Integrated Design and Analysis (2013)

Paul, J.: Reinforced soil systems in embankments construction practices. In: Proceedings of International Geotechnical Symposium on Theory and Practice of Earth Reinforcement, Fukuoka, Japan, October, pp. 461–466 (1988)

Pokharel S.K., et al.: Experimental study on bearing capacity of geocell-reinforced bases. In: 8th International Conference on Bearing Capacity of Roads, Railways and Airfields, June 29–July 2, Champaign, Illinois (2009)

Pokharel, S.K., et al.: Investigation of factors influencing behaviour of single geocell-reinforced bases under static loading. Geotext. Geomembr. Elsevier (2010). https://doi.org/10.1016/j.geotexmem.2010.06.002

Rajagopal K., et al.: Behaviour of sand confined with single and multiple geocells. Geotext. Geomembr. Elsevier (1999). https://doi.org/10.1016/s0266-1144(98)00034-x

Rea, C., Mitchell, J.K.: Sand Reinforcement Using Paper Grid Cells, pp. 644–663. Regular meeting-Rocky Mountain Coal Mining Institute (1978)

Robertson, J., Gilchrist, A.J.T.: Design and construction of a reinforced embankment across soft lakebed deposits. In: Proceedings of International Conference on Foundations and Tunnels, pp. 84–103. London, 2, M C Forde Engineering Technics Press, Edinburg (1987)

Sitharam, T.G., Hegde, A.: Design and construction of geocell foundation to support the embankment on settled red mud. Geotext. Geomembr. Elsevier (2013). https://doi.org/10.1016/j.geotexmem.2013.08.005

Sowmiya, L.S.: Analyses and Experimental Investigations of Railway Tracks with and without Geosynthetic Reinforcement. Ph.D. Thesis, Indian Institute of Technology, Delhi, India (2013)

Tafreshi, S.N.M., Dawson, A.R.: A comparison of static and cyclic loading responses of foundations on geocell-reinforced sand. Geotext. Geomembr. Elsevier (2012). https://doi.org/10.1016/j.geotexmem.2011.12.003

Zhang, L., et al.: Bearing capacity of geocell reinforcement in embankment engineering. Geotext. Geomembr. Elsevier (2010). https://doi.org/10.1016/j.geotexmem.2009.12.011

A Study on Utilization of Mine Overburden as a Replacement of Base and Sub-base Layers on Rural Roads

Brajendra Nath, Sowmiya Chawla(✉), and Randhir Kumar Gupta

Department of Civil Engineering, IIT(ISM), Dhanbad, Jharkhand, India
pranav.talkto@gmail.com, sowmiya_iitd@hotmail.com, randhir@live.in

Abstract. The ever-increasing urbanization, globalization and population have led to construction of pavement in rural areas. It has been a cause of concern of depletion of natural resources at an alarming rate but also gradually becoming a challenge for sustainability. This paper presents the application of mine overburden (OB) dump material for replacing the natural aggregate in rural roads. OB dump also known as colliery spoil which is the solid residual material resulting from the mining of coal. It is likely to contain varying proportions of sandstone, shale, mudstone and coal fragments. These overburden can be used in various ways as a substitute for conventional material for rural roads. The coal producing countries (including India) accumulate large amounts of OB close to collieries. The existing coal mine overburden might be a good alternative for aggregate and provide a suitable base and sub base material in rural roads. Use of OB may provide an economical solution to reduce the transportation cost that has to be hauled from a long distance under such conditions also reduce the environmental influence caused by coal mining. In this present study the 1:3 scaled laboratory model tests have been carried out on an unpaved pavements with OB as base and sub-base layers. Monotonic as well as cyclic tests have been conducted using computer controlled pneumatic cyclic actuator. The tests were conducted with various soil conditions and loading patterns. A finite element analysis for monotonic loading has been done with different soil conditions. Based on the test results, it can be concluded that Dhanbad soil of subgrade CBR 3 gives better results than other subgrade of low CBR. The overburden aggregate shows similar behaviour to that of virgin aggregates. Therefore overburden dumps can be used as a replacement of base and sub-base layers for rural roads where the utility of low volume roads are high.

1 Introduction

Rural Roads are recognized as an infrastructure for the social, economic and agricultural growth of the country. Rural infrastructure has a direct link in improving livelihoods, health and productivity and reduced poverty. Rural roads constitute more than 80% of the total road network in India Fig. 1 and Table 1. Roads are a lifeline for rural communities, linking them to markets, education, health and other facilities. Better

M.-C. Weng et al. (eds.), *Current Geotechnical Engineering Aspects of Civil Infrastructures*, Sustainable Civil Infrastructures,
https://doi.org/10.1007/978-3-319-95750-0_7

roads provide improved market access which, in term results in favorable input and output prices and improving the economic condition of the rural poor.

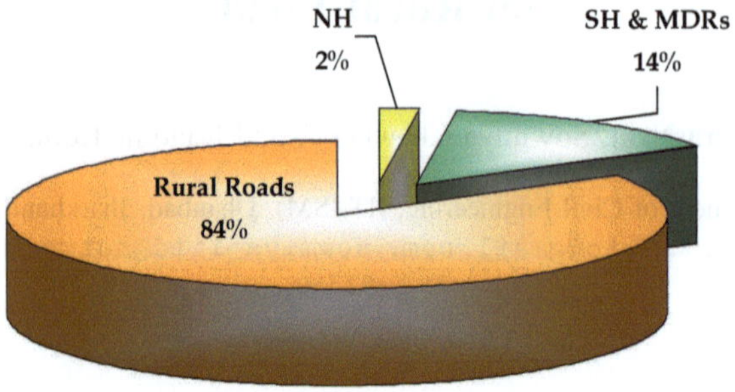

Fig. 1. Status of rural roads in India

The Government of India in order to accelerate development and growth made huge investments in power, industry, transport and essential social services like education and health. Efforts were also made to improve the productivity of agriculture sector in the country. However benefits of all these investments were not reaching the majority of its population living in the rural areas. In the early 70s, planners decided to give a major thrust to the development of rural roads through investing in various schemes which had provisions to provide help for developing rural roads despite of that the rural roads development still lacked adequate planning and management and the need for National and State highways remained a priority in the successive years. To improve and maintain its usage alternative such as use of mine stone overburden in low volume and unpaved roads need to be explored. As most of the roads in the United States and the world are low-volume unpaved roads (Tingle and Jersey 2007), materials required for construction and maintenance of these roads put a huge burden on the environment every year. The possible use of mine stone overburden to replace natural and produced materials therefore could provide a better choice in environmental terms as compared to other materials such as sand, gravel and aggregate.

1.1 Production of Mine Waste

Coal is the most abundant fossil fuel resource present in India. Currently, India stands fourth in the world in terms of identified reserves. There are 44 major coalfields located in the Indian peninsular, in addition to 17 in the north-eastern region. In India production from opencast mining which currently accounts for 73.8% of total coal production. India has reached the forefront of the world coal scene, ranks fourth in total coal production and is the third largest coal producer from opencast mines. The average stripping ratio (overburden to coal) during the last three decades was 1.97 tonnes/m^3 (Chalukya 2000). Increasing coal mining waste and the accelerating pace of industrialization have stretched the self-sustaining capacity of the ecosystem to the point where its delicate equilibrium is threatened. Table 1 and Fig. 2 gives a clear cut idea about the production of mine stone or OB dumps in worldwide and in India during the mining process.

Table 1. Total length of road network in India 32 lakh km

S.no	Road network in India	Length (km)
1	National highways (NH)	65,600 km (2.04%)
2	State highways and major district roads (SH and MDRs)	4,32,000 km (13.5%)
3	Rural roads	27 lakh km (84.4%)

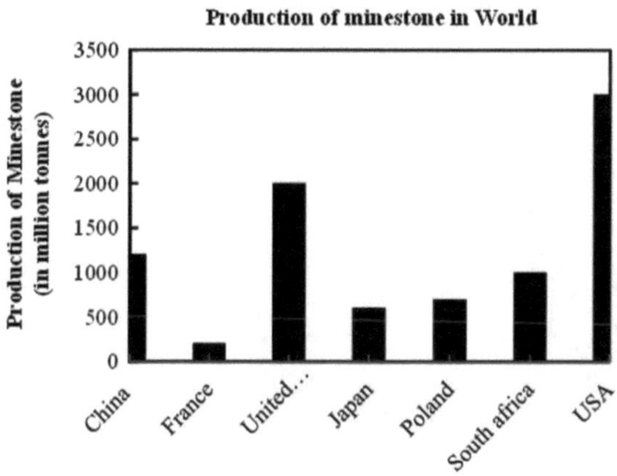

Fig. 2. Production of mine stone in the world (after Krystyna M Skarzynska 1995)

Opencast excavation of coal deposits involves the removal of overlying soil and rock debris and their storage in overburden dumps. These overburden dumps change the natural land topography, affect the drainage system and prevent natural succession of plant growth resulting in acute problems of soil erosion and environmental pollution. The overburden consists of alluvial loose sand, gravel, shale and sandstone. Sandstone are sedimentary rocks consists of quartz and feldspar and found in various colours like white, grey, red, buff, brown, yellow and even dark grey. The specific gravity of sandstone varies from 1.85 to 2.7 and compressive strength varies from 20 to 170 N/mm^2. Its porosity varies from 5 to 25% (Sharzynska 1995a). There is various application of mine stone in railways, roadways, harbour, mining and hydraulic Engineering as described in (Sharzynska 1995b). The results on test sections with other base materials (sand, aggregate and quarry waste) can be found in Pokharel et al. (2010, 2011), and Yang et al. (2011).

2 Facility, Equipment and Test Preparation

Test facilities at the Indian Institute of technology ISM Dhanbad were used for this research. The material properties were determined at IIT ISM Dhanbad. A photographic view of the model test set-up is given in Fig. 3. To apply monotonic and cyclic loadings on the model, the tank was mounted on the loading frame of a cyclic Actuator with Computer Control System. The set up consists of specially fabricated rectangular steel tank with the dimension of 120 cm × 80 cm × 60 cm and is made up of 6 mm thick steel plates and suitable angles acting as stiffeners for conducting model tests on low volume roads. The tank is fitted with Perspex sheet on both sides so that observers could visualize the failure of the pavement. The other two sides of the tank are smooth and rigid to create plane strain conditions in the tank. A Polythene sheet lubricated with oil on the wall to reduce friction and allow free vertical movement. The granular layer (sub base and base layer) and subgrade were compacted using a 10 kg steel hammer and a 20 kg specially manufactured sheep foot roller respectively.

Fig. 3. The typical model test

2.1 Instrumentation Used

Monotonic and cyclic loading were applied at the top of the pavement by a circular plate which was in turn connected to pneumatic load cell of the cyclic actuator. Model pavement were adequately instrumented to record important responses during the tests. The instrumentation consists of displacement transducer and pneumatic load cell to a computer controlled data acquisition system. A load cell of 20 kN capacity was used to measure the vertical load.

3 Test Procedure

3.1 Bed Preparation and Instrumentation

The subgrade soil bed was prepared as follows: first, the dry soil was sieved to remove foreign particles; a required amount of water was then mixed with the soil. The soil which was used in this study namely Dhanbad soil was mixed with water manually by bare feet in a 30-kg batch each in a large ceramic tub to break up any clods present base on a trial and error procedure. The soil was then filled up in separate cylindrical drums and covered up with wet jute bags for 24 h for moisture equalization. At the bottom of the test tank, circular holes were provided at constant interval. These holes were covered with a layer of non-woven geotextile and subgrade soil was placed over it ensuring a drainage boundary at the bottom of the tank. The test tank was filled with required amount of subgrade soil and compacted in four layers to achieve desired density. Kneading compaction was carried out using a small specially manufactured sheep foot roller of 20 kg weight as shown in Fig. 4. After the compaction process samples were collected from three different location in plan and from four different depths at each location to monitor water content, dry density and undrained shear strength. The entire surface area was then covered with wet jute sheets and the soil bed was left in this condition for further 24-hour moisture equalization. The sub-base and base layer which consist of overburden aggregates were prepared and compacted in two layers by manual compaction to ensure uniformity. The required amount of moorum was added in granular layer to fill the gap between the aggregates.

Fig. 4. Kneading compaction using sheep foot roller

3.2 Loading System

The axle load in most states varies from 80 kN (18,000 lb) to 90 kN (20,000 lb) (Yoder and Witczak 1975). In this study, 1:3 scaled laboratory model test consist of single tire of 20 kN (4500 lb) is considered. The load is transferred to the pavement surface through a contact pressure of single tire. The tire contact pressure on the road is equal to the tire pressure, which is equivalent to 550 kPa. A circular loading plate was used to apply the wheel load on the laboratory model pavement section. The following equation were used to determine the dimensions of the loading plate to simulate the effect of a wheel load (Yoder and witczak 1975):

$$r_c = \sqrt{\frac{P}{P_c \pi}} \quad (1)$$

$$D = 2r_c \quad (2)$$

where, r_c is the radius of contact surface, P is the total load on the tire, P_c is the tire inflation pressure and D is the diameter of contact surface. In this study the thickness of the loading pate was 20 mm and the diameter of the loading plate was 124 mm.

3.3 Model Test Procedure

A total of 8 tests were conducted in two groups on pavement section. Group 1 was termed as monotonic test and Group 2 as Cyclic test. The pavement section consisted of Dhanbad soil as subgrade of depth 40 cm, overburden aggregate as sub-base and base of depth 15 cm and 5 cm respectively. The designed pavement section was according to the guidelines for the design of flexible pavements for rural roads (IRC SP: 72-2007). The above section was in a reduced laboratory model test of 1:3 scale. In Group 1 as termed as monotonic test, four tests were conducted on different soil condition of Dhanbad soil and in Group 2, four cyclic tests were conducted on subgrade of Dhanbad soil of CBR 2 of load 2.4 and 3.6 kN and CBR3 of load 2.4 and 3.6 kN.

4 Material Properties

4.1 Subgrade

The locally available Dhanbad soil is used as a subgrade soil for this study which are classified as sandy silt of low plasticity. Important characteristics of the Dhanbad soil are summarized in Table 2. Standard Procter compaction and CBR tests on this soil were carried out in the Laboratory. The optimum moisture content and the maximum dry density were 16% and 18.8 kN/m^3. The compaction curves are shown in Fig. 5. The grain size distribution of subgrade as shown in Fig. 6 has a D_{10} of about 0.18 mm, a D_{60} of about 0.82 mm and a uniformity coefficient of about 4.55. It can be concluded that soil is well graded soil (Table 3).

Table 2. Overburden of mining waste in India (in million tonnes)

	1999–2000	2000–2001	2001–2002	2002–2003	2003–2004	2004–2005	2005–2006
1. Coal							
Production	300	310	323	337	356	377	407
overburden	1100	1135	1183	1235	1304	1383	1493
2. Bauxite							
production	7.1	7.99	8.59	9.87	10.92	11.96	12.34
overburden	4.3	4.84	5.20	6.0	6.6	7.2	7.5
3. Limestone							
Production	129	123.6	129.3	155.74	153.39	165.75	170.38
overburden	135	129.4	135.3	163.0	160.5	173.5	178.3
4. Iron ore							
Production	75	80.7	86.2	99.1	122.8	145.9	154.4
overburden	69.9	75.2	80.3	92.3	114.5	136	143.9
5. Others							
Production	7.5	5.02	4.91	7.82	8.32	9.24	9.44
Overburden	14.5	10.60	10.38	15.20	16.26	19.30	18.61

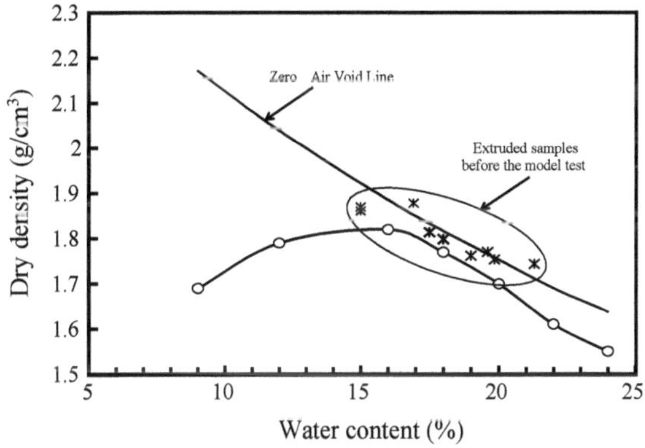

Fig. 5. Water content and dry density relationship of Dhanbad soil

5 Mine Overburden Aggregates

Mine overburden material named as sandstone brought from a mine overburden dump site in Jharia coalfield Dhanbad, Jharkhand, were used in this study as sub-base and base layers. Materials used for sub-base and base layer was a crushed aggregate of sandstone of sedimentary rock. The crushed aggregates are then sieved in reduced scale of 1:3 of gradation of sub base and base layer according to the code of rural roads manual (IRC SP:20-2002). The D_{10} and D_{60} sizes are approximately 11 and 28 mm respectively, giving a uniformity coefficient of 2.54. Therefore sub base can be

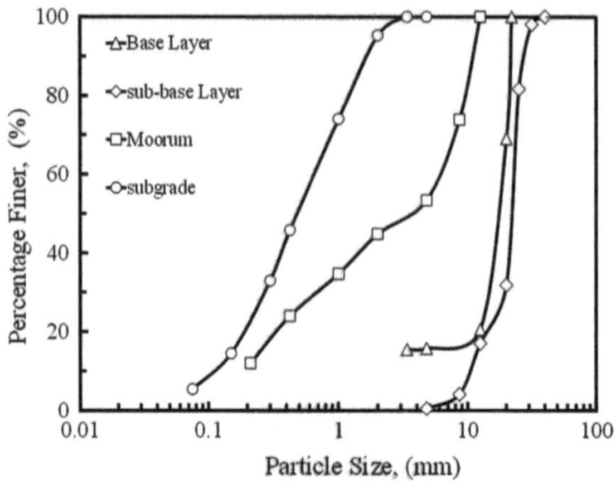

Fig. 6. Grain size distribution of various pavement layer materials

Table 3. Physical properties of subgrade soil

S. no	Properties	Dhanbad soil
1	Dry density (kN/m^3)	18.8
2	OMC (%)	16
3	Specific gravity	2.62
4	Coefficient of uniformity, C_u	4.55
5	Coefficient of curvature, C_c	0.978
6	D_{10} (mm)	0.18
7	D_{30} (mm)	0.38
8	D_{60} (mm)	0.82

classified as a uniform fine gravel. The specific gravity, water absorption, aggregate impact test, los angles abrasion test and slake durability were conducted as shown in Table 2. Aggregate impact value indicates a relative measure of the resistance of aggregate to impact. Bureau of Indian standards (BIS) for aggregate impact value are used to classify the stone aggregates with respect to toughness property. The aggregate impact value of soft aggregates for road construction is given in Table 4.

Table 4. Basic properties of mine stone aggregate

S. no	Properties	Minestone	Physical requirement of course aggregate for WBM for sub base and base course (IRC)
1	Specific gravity	2.28	2.5–3.2
2	Water absorption (%)	4.7	4.0
3	Aggregate impact test (AIV) (%)	36	50
4	Los Angeles abrasion test (%)	69	40
5	Slake durability test (%)	96	–

6 Finite Element Method

Finite Element Method (FEM) is a powerful tool to analyze engineering problems, the accuracy of which depends on several factors. First, the extent of meshing of the individual components. Second, incorporation of the parameters that has major influence on the problem. Finally, the interpretations of the result should be reasonable. Accordingly, the FEM is used in this study to perform a comparative analysis of the results obtained from the two methods i.e. experimental and FEM.

6.1 Material Properties

Material properties used for the FEM analysis is same as that for the experimental setup. The properties of each component is shown in Table 5.

Table 5. Material properties

	Soil layer	E (kPa)	μ	Υ (kN/m^3)	C	ϕ
CBR < 0.16	Subgrade	100	0.49	18	4	0.1
	Sub-base	190	0.35	20	0.1	35
	Base	240	0.3	21	0.1	30
CBR = 0.2	Subgrade	100	0.49	18	4	0.1
	Sub-base	3800	0.35	20	0.1	35
	Base	5200	0.3	21	0.1	30

6.2 Finite Element Model

Finite Element Modelling of the experimental setup is done using the geotechnical analysis software, Midas GTS NX. A three-dimensional model of the subgrade, sub-base and base layers is loaded on top to get the load settlement profile of the base layer. The subgrade is modelled using elastic-plastic, Mohr-Coulomb constitutive relationship, the sub-base and base layers are modelled using elastic-plastic, Mohr-coulomb, effective stress (drained) relationship. Each layers of pavement are divided into proper element size using the auto mesh option. The FEM model is shown in Fig. 7.

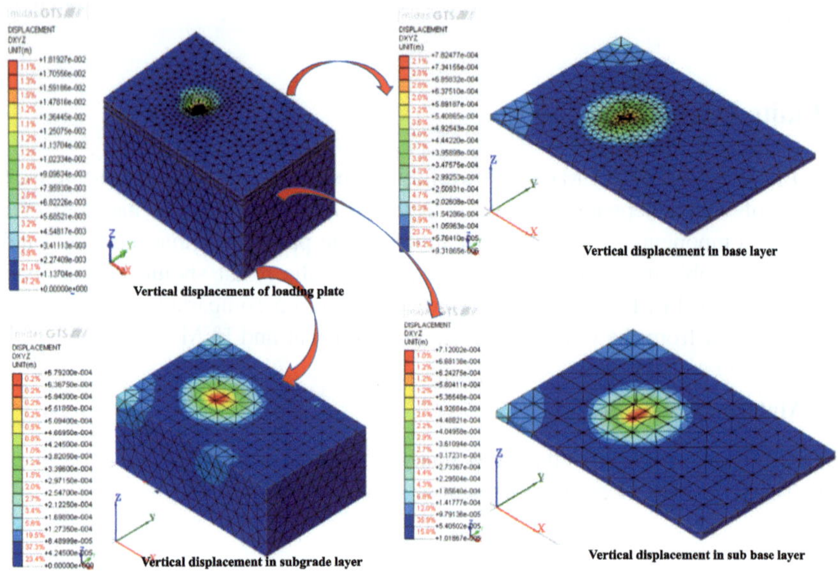

Fig. 7. Finite Element method test

6.2.1 Boundary Conditions

The use of proper boundary condition is the most vital part so as to represent the exact physical setting of the actual problem. Therefore, to obtain this several tools are used such as restraints to the nodes of boundary layer. In this study, roller supports is used for the vertical boundaries and the bottom boundary face is considered as fixed. The intermediate boundaries between the layers are set as drainage boundary.

6.2.2 Loading

A vertical load of 10 kN is applied on the plate with 10 load steps with a maximum of 350 iterations in each step. The loading is performed for twenty minutes and displacements for each loading step is obtained.

7 Results and Discussions

7.1 Monotonic Tests

A strain controlled monotonic test was conducted on an unreinforced pavement section. The result of this test was used to determine the magnitude of cyclic load to be applied during cyclic tests. During the monotonic test the loading was applied at the strain rate of 0.06 mm/sec. These tests were conducted to find out the effect of subgrade condition on the load settlement responses. Based on (Vesic 1963) the failure loads under monotonic loading for different soil conditions for unreinforced pavements were obtained as 4 kN and 6 kN respectively. A threshold stress is defined as critical stress level is above CLRS, plastic deformation is non-terminating. Based on (Sharma 2007 and Sowmiya 2013) threshold stress ratio (TSR) has been obtained as 60% for low plasticity soils. Based on this cyclic load of 2.4 kN (TSR = 60%) of subgrade of CBR 2 and 3.6 kN (TSR = 60%) of subgrade of CBR 3 were used for different soil conditions of Dhanbad soil.

7.1.1 Effect of Subgrade CBR on Total Settlement

Figure 8 presents the effect of Subgrade soil CBR value on the total settlement of the pavement. From the results it is concluded that CBR of less than 2 shows the behavior of punching shear failure and local shear failure and CBR above 2 shows General shear failure.

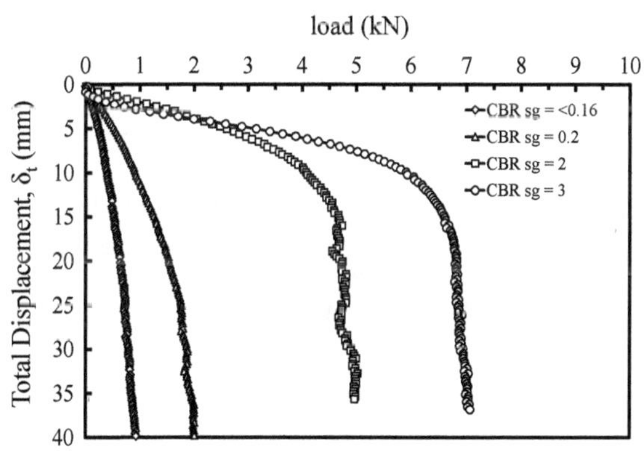

Fig. 8. Effect of subgrade soil CBR value on the load—settlement response

7.1.2 Effect of Compaction Energy on Total Displacement

Figure 9 presents the effect of compaction effort on the total displacement. It is concluded that the subgrade compacted with sheep foot roller gives better load carrying capacity than the subgrade compacted with hand kneading.

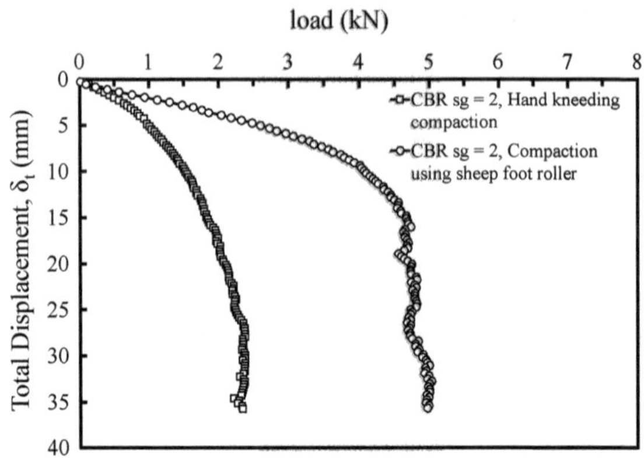

Fig. 9. Effect of compaction effort on the load—settlement response

7.1.3 Finite Element Prediction

Figure 10 presents the finite Element prediction of the comparison of load displacement curves of model test and numerical model. Both the model test and numerical test shows the similar behavior compacted with hand kneading and sheep foot roller. The compaction with hand kneading shows that soil is loosely packed and shows punching and local shear failure while compaction with sheep foot roller shows that soil is capable of taking the desired load for transportation with nature of General shear failure.

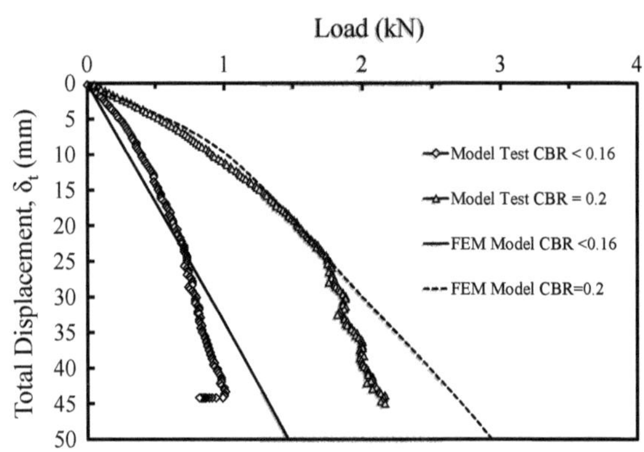

Fig. 10. Comparison of displacement in model test and FEM model

7.2 Cyclic Tests

The cyclic tests of pavement section was carried out on two different soil conditions of CBR 2 and CBR 3 of load 2.4 and 3.6 kN. For the subgrade of CBR 2 threshold stress ratio was taken as 60% as 2.4 kN of 4 kN failure load and cyclic tests were conducted in four steps i.e. 1.2 kN (30% of failure load), 1.6 kN (40% of failure load), 2 kN (50% of failure load) and 2.4 kN (60% of failure load). Considering the Indian typical mixed traffic condition varying axle loads in a particular lane, the above load steps have been applied. For the subgrade of CBR 3 threshold stress ratio was taken as 60% as 3.6 kN of 6 kN failure load and cyclic tests were done in four steps i.e. 1.8 kN (30% of failure load), 2.4 kN(40% of failure load),3 kN(40% of failure load) and 3.6 kN (60% of failure load).

7.2.1 Loading Pattern of Cyclic Model Test

The actuator with load cell was placed over loading plate. A sinusoidal load pattern has been followed at an increasing magnitude of load cycle of 100 to 25,000 with frequency of 0.67 Hz (Perkins et al. 2012) has been applied with Dhanbad soil as subgrade with different soil conditions of 2.4 and 3.6 kN cyclic load was applied as shown in Figs. 11 and 12. Initially the stiffness of the model pavement was very low for low volume roads, therefore a slow frequency of 0.67 Hz was applied. A strain controlled loading was applied during monotonic tests while a stress controlled loading was applied during cyclic tests.

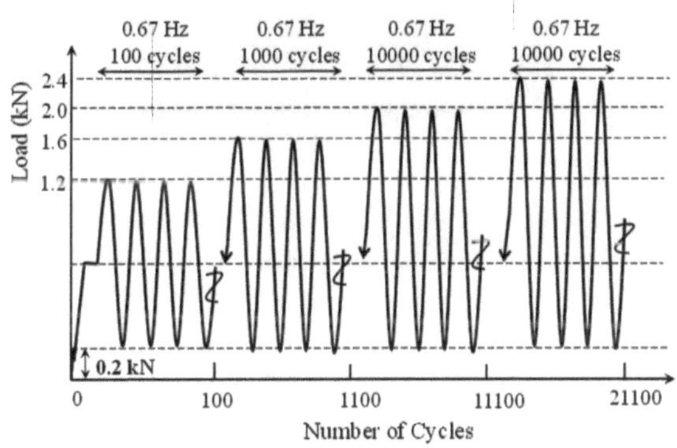

Fig. 11. Sinusoidal loading sequence of 2.4 kN

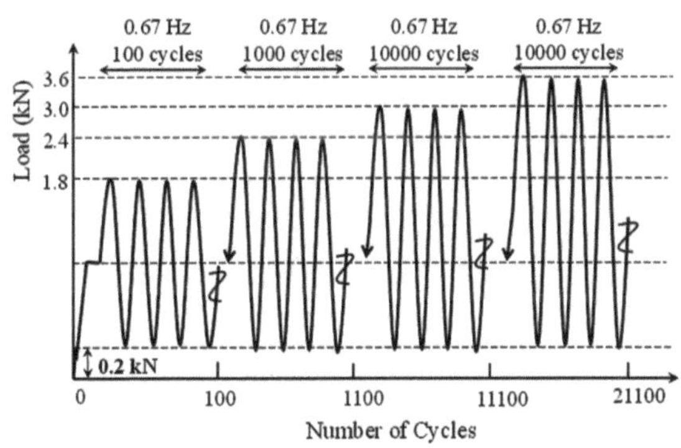

Fig. 12. Sinusoidal loading sequence of 3.6 kN

7.2.2 Effect of Subgrade CBR on Rut Depth

Figure 13 shows the typical initial cyclic loading pattern up to 1000 loading cycles of the model test results of CBR 3 with 3.6 kN. Figure 14 shows the model test section, the effect of subgrade CBR 2 and 3 on rut depth for cyclic load of 2.4 kN. The pavement section with subgrade CBR 2 shows more rut depth as compared to subgrade CBR 3 for different steps of loading pattern of 1.2, 1.6, 2.0, 2.4 kN respectively. It is due to the fact that CBR 3 is stiffer than CBR 2.The behavior of subgrade of CBR 3 shows the pattern of general shear failure while the CBR 2 shows the pattern of local shear failure.

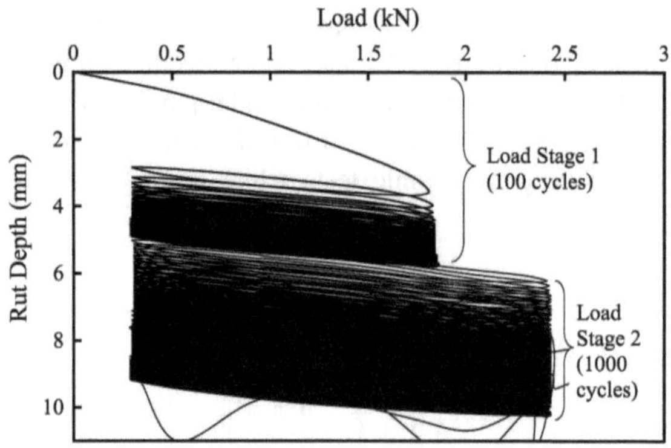

Fig. 13. Typical initial cyclic loading pattern up to 1000 loading cycles of the model test results of CBR 3 with 3.6 kN

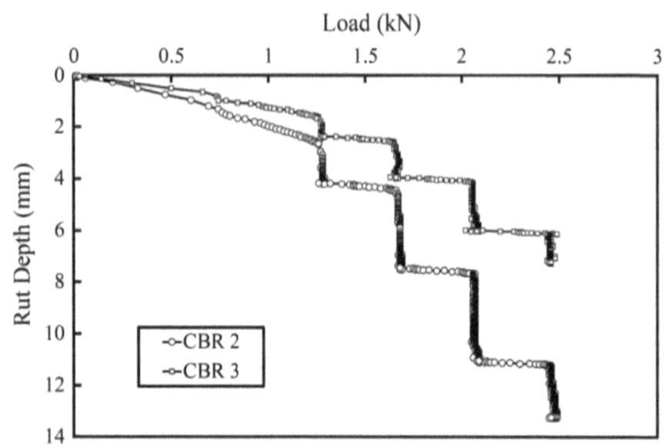

Fig. 14. Effect of subgrade CBR on rut depth for cyclic load of 2.4 kN

Figure 15 shows the model test section the effect of subgrade CBR 2 and 3 on rut depth for cyclic load of 3.6 kN.The pavement section with subgrade CBR 2 shows more rut depth as compared to subgrade CBR 3 for different steps of loading pattern of

1.8, 2.4, 3.0, 3.6 kN respectively. It is due to the fact that CBR 3 is stiffer than CBR 2. It is concluded that the behavior of subgrade of CBR 3 shows the pattern of general shear failure while the CBR 2 shows the pattern of local shear failure.

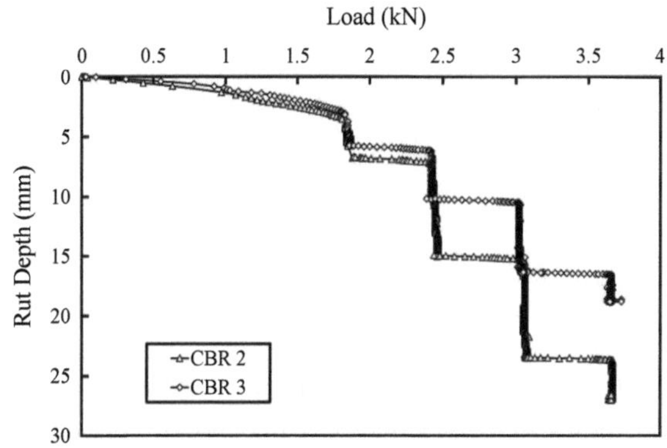

Fig. 15. Effect of subgrade CBR on rut depth for cyclic load of 3.6 kN

7.2.3 Effect of Cyclic Loads on Subgrade CBR

Figures 16 and 17 shows the effects of cyclic loads of 2.4 and 3.6 kN on subgrade CBR 2 and 3, respectively. In seems that, with same subgrade CBR 2 and different loading conditions the displacement is more in case of cyclic load of 3.6 kN than 2.4 kN. For the first 100 and 1000 cycles the rut depth is steep in both the cases. In third cycle of 10,000 the rut depth is increasing but constant in both the loading cases of same soil condition. In fourth cycle of 10,000, obviously the rut depth is increasing but constant in both the cases. The modulus of the pavement section has been changed due to increasing traffic loading. After 'n' number of traffic passes the pavement layers get compacted and the overall stiffness of the pavement section also increased. This is reason for the reduction in the incrimination of the rut depth with increasing the loading cycles.

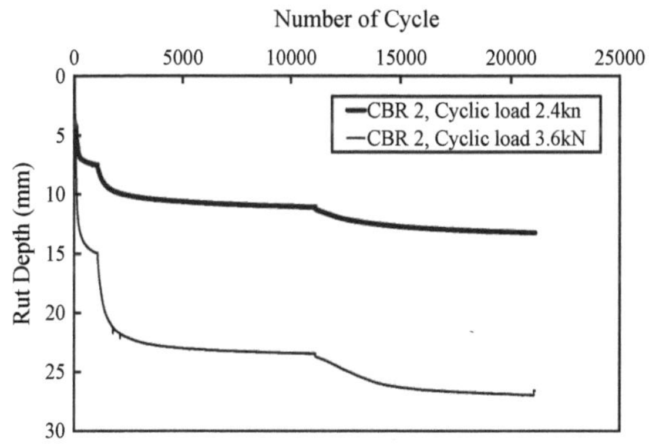

Fig. 16. Effect of cyclic loads of 2.4 and 3.6 kN on subgrade CBR 2

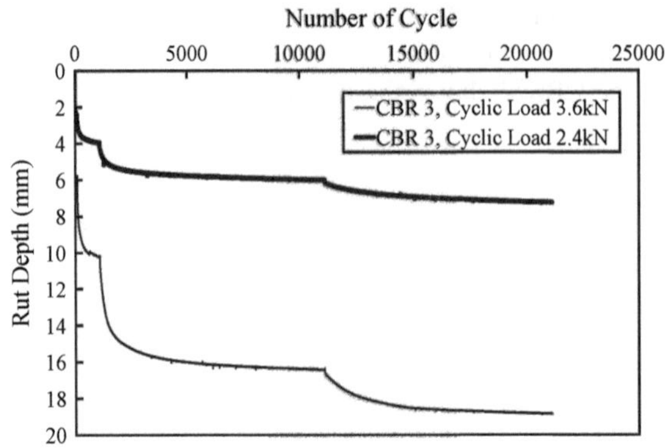

Fig. 17. Effect of cyclic loads of 2.4 and 3.6 kN on subgrade CBR 3

Figures 18 and 19 shows the effect of cyclic load 2.4, 3.6 kN on subgrade of CBR 2 and 3, respectively. For the first 100 cycles the rut depth is more in subgrade with CBR 2 than subgrade with CBR 3. For the second 1000 cycles the rut depth is steep in both the soil conditions. For the third 10,000 cycles rut depth is constant in both the subgrade of CBR 2 and CBR 3 but rut depth has increased significantly in subgrade of CBR 2. For the last 10,000 cycles rut depth is uniform in both the cases.

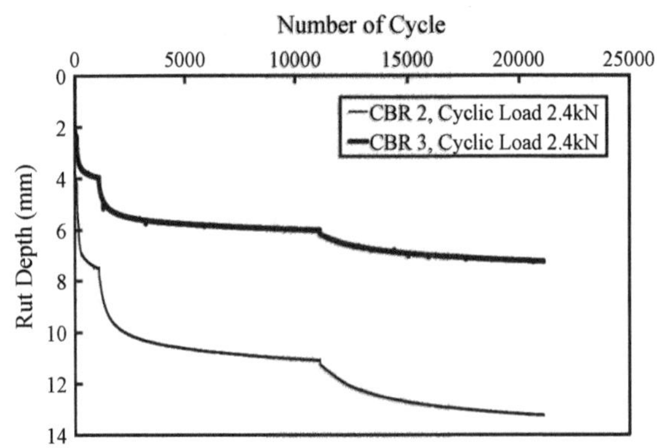

Fig. 18. Effect of cyclic load 2.4 kN on subgrade CBR of 2 and 3

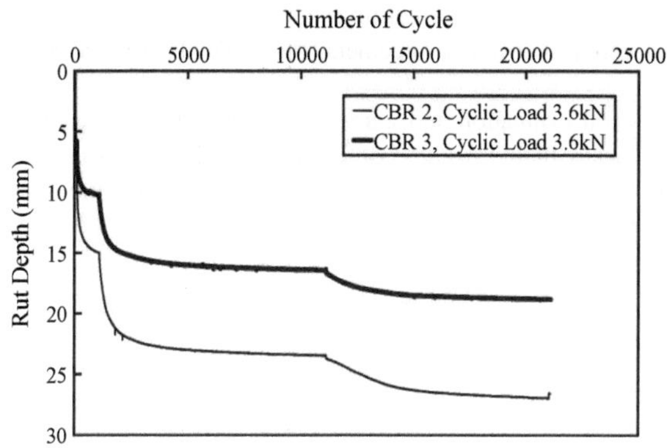

Fig. 19. Effect of cyclic load 3.6 kN on subgrade CBR of 2 and 3

8 Conclusion

This paper presents the results of the experiment conducted on reduced scale model test and numerical tests on sample test section that include mine stone overburden materials in base and sub base layer. The following conclusion can be drawn from this study that Mine stone overburden materials can be used as pavement materials for base and sub base courses in unpaved roads as sustainable alternative. Laboratory model test results shows that the subgrade condition will play a major role for the load settlement response of the pavement. This is the preliminary study of the behaviour of modelled rural pavement using overburden aggregates as a base as well as sub-base layer. Moreover, the detailed study has been needed to study the resilient behaviour as well as the breakage characteristics of the overburden aggregates.

References

Chaulya, S.K.: Bioreclaimation of coal mine overburden dumps in India. Land Reclaimation Contam. **8**(3), 2000 (2000)

IRC SP: 72-2007 Guidelines for the design of flexible pavements for Low volume rural roads

IRC: SP: 20-2002 Rural roads manual (2002)

Perkins, S.W., Christopher, B.R., Lacina, B.A., Klompmaker, J.: Mechanistic-empirical modeling of geosynthetic reinforced unpaved roads. Int. J. Geomech. **12**(4), 370–380 (2012)

Pokharel, S.K., Han, J., Leshchinsky, D., Parsons, R.L., Halahmi, I.: Investigation of factors influencing behavior of single geocellreinforced bases under static loading. Geotext. Geomembr. **28**(6), 570–578 (2010)

Pokharel, S.K., Han, J., Manandhar, C., Yang, X., Leshchinsky, D., Halahmi, I., Parsons, R.L.: Accelerated pavement testing of geocell-reinforced unpaved roads over weak subgrade. J. Transp. Res. Board, Low-Volume Roads **2**(2204), 67–75 (2011)

Skarzynska, K.M.: Reuse of coal mining wastes in civil engineering part. Waste Manage. **15**(2), 83–126 (1995a)

Skarzynska, K.M.: Reuse of coal mining wastes in civil engineering: properties of minestone. Waste Manage. **15**(1), 3–42 (1995b)

Sowmiya, L.S.: Analysis and Experimental investigation of railway tracks on clayey subgrades with and without geosynthetics reinforcement. Ph.D. Thesis, Indian institute of Technology, Delhi, India (2013)

Sharma, A.: Geotechnical evaluation and numerical modelling of railway tracks on compacted subgrade. Ph.D. Thesis, Indian institute of Technology, Delhi, India (2007)

Tingle, J.S., Jersey, S.R.: Empirical design methods for geosynthetic-reinforced low-volume roads. Transportation Research Record 1989, vol. 2, pp. 91–101. Transportation Research Board, Washington, DC (2007)

Vesic, A.S.: Bearing capacity of deep foundations in sands. National Acad. Sci. National Res. Counc. Highw. Res. Rec. **39**, 112–153 (1963)

Yoder, E.J., Witczak, M.W.: Principles of Pavement Design, 2nd edn. Wiley, New-york (1975)

Yang, X., Han, J., Pokharel, S.K., Manandhar, C., Parsons, R.L., Leshchinsky, D., Halahmi, I.: Accelerated Pavement Testing of Unpaved Roads with Geocell-Reinforced Sand Bases. Geotext, Geomembrane (2011)

A New Equation for SWCCs of Unsaturated Soils

Suched Likitlersuang[1(✉)] and Martin D. Liu[2]

[1] Geotechnical Research Unit, Department of Civil Engineering, Faculty of Engineering, Chulalongkorn University, Bangkok, Thailand
fceslk@eng.chula.ac.th
[2] School of Civil Mining and Environmental Engineering, Wollongong University, Wollongong, Australia
martindl@uow.edu.au

Abstract. In the current geotechnical engineering practice, soil-water characteristic curves (SWCCs) are essential for quantifying the mechanical responses of unsaturated soil. The SWCC can be measured from laboratory or estimated from some empirical equations. Measured SWCCs are usually conducted for drying curve by using Tempe cell or pressure plate device. Because of the time consuming associate the absorption, the wetting SWCC can be approximated by estimating the magnitude of the hysteresis loop at the inflection point on the drying SWCC. The SWCC is commonly plotted as a curve between (gravimetric or volumetric) water content and and the logarithm of soil suction. The general shape of the SWCC looks similar to the *S*-shaped function. An accurate mathematical modelling of the S-shaped curve is usually required for engineering computation such as numerical analyses. In this paper, a new method to quantifying the SWCC, i.e., via the S-shaped equation proposed by Liu et al. (2013), is employed to represent the SWCC. Comparisons between the equation simulations and experimental data are made. Discussions on the equation simulations and characteristics of the mechanical properties of unsaturated soils are also presented.

1 Introduction

Soil-water characteristic curves (SWCCs) have an important role for unsaturated soil, as it can be used to estimate many property functions such as permeability and shear strength functions. The SWCC is commonly plotted as a curve between (gravimetric or volumetric) water content and the logarithm of soil suction as shown in Fig. 1. The general shape of the SWCC looks similar to the *S*-shape function, in which two points of change in slope along the curve have physical meanings. The first point is defined as air-entry value, where the largest voids start to desaturate as suction is increased. The second point is defined as residual condition, where the removal of water from the soil becomes considerably difficult. The SWCC can be then divided into three zones, namely, the boundary effect zone in the low suction range, the transition zone between the air-entry value and the residual value, and the residual zone at high suctions as depicted in Fig. 1. The SWCC is not a single-valued curve, but it is hysteretic nature

M.-C. Weng et al. (eds.), *Current Geotechnical Engineering Aspects of Civil Infrastructures*, Sustainable Civil Infrastructures,
https://doi.org/10.1007/978-3-319-95750-0_8

associated with the drying (desorption) and wetting (absorption) as presented in Fig. 2. Consequently, it is difficult to know whether the unsaturated soil is presently on the drying curve, the wetting curve or somewhere in between these two boundaries (Fredlund et al. 2012).

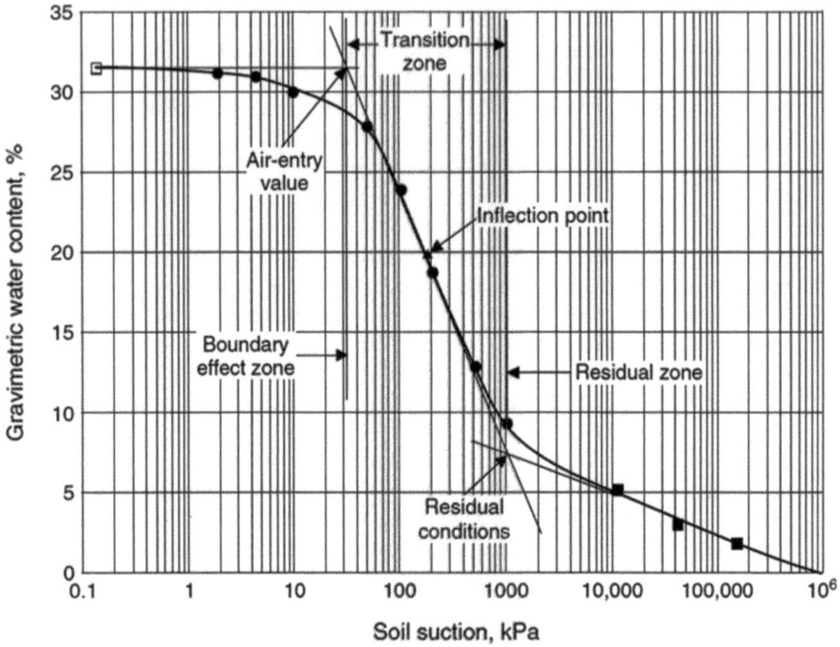

Fig. 1. Typical SWCC (after, Frendlund et al. 2012)

In geotechnical engineering practice, the SWCC can be measured from laboratory or estimated from some empirical equations. Measured SWCC are usually conducted for drying curve by using Tempe cell or pressure plate device. On the other hand, the laboratory tests for measuring wetting curve are considered too costly because of the time consuming associate the absorption. The wetting SWCC can be approximated by estimating the magnitude of the hysteresis loop at the inflection point on the drying SWCC. For simplicity reasons, the effect of hysteresis can be sometime ignored and use only the drying SWCC for the estimation. (i.e., the upper bounding branch of the SWCCs shown in Fig. 2).

There are several empirical equations that have been proposed to estimate SWCCs (i.e., Brooks and Corey 1964; van Genuchten 1980; Fredlund and Xing 1994 and Fredlund and Pham 2006). Most of SWCC equations present the form of a continuous function being asymptotic at both ends. Only the transition zone between the air-entry value and residual suction has sufficient slope for the calculation of soil suction. Fredlund et al. (2011) summarized some commonly used SWCC equations including more recent development in geotechnical engineering. The SWCC equations can be categorized based upon the range of soil suction conditions that can be well defined by the equations. It is noted that the proposed SWCC equations have been firstly developed and applied in agriculture-related disciplines. However, some mathematics and their physical meanings in geotechnical and geoenvironmental engineering are somewhat different from those in agriculture.

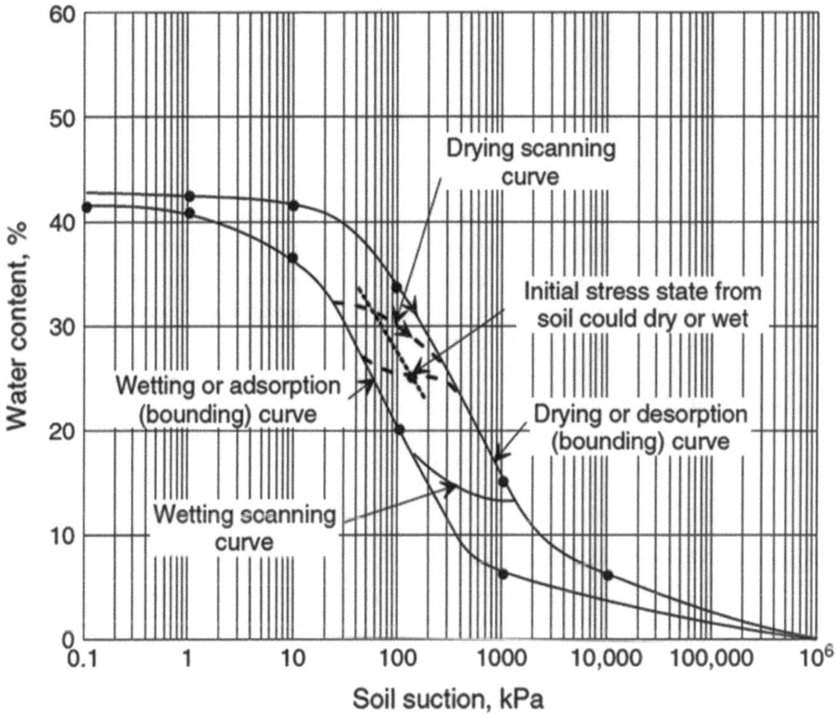

Fig. 2. Hysteresis loops comprising the SWCC (after, Fredlund et al. 2012)

It has been widely observed that the SWCC is *S*-shaped. Based on the mathematical function proposed by Liu et al. (2013), a new method/equation is proposed to describe the SWCC in this study. The SWCCs for different soils under various conditions are simulated, and comparisons between the equation simulations and experimental data are made. Discussions on the equation simulation and characteristics of the mechanical properties of unsaturated soils are also presented.

2 Proposed Equation for SWCCS

Because of the important role of SWCCs in unsaturated soil mechanics, several equations for SWCCs have been proposed and used in research and engineering practice. For examples, they are Brooks-Corey equation (Brooks and Corey 1964), van Genuchten equation (van Genuchten 1980), and Fredlund and Xing equation (Fredlund and Xing 1994). There are four parameters for Brooks-Corey equation and for van Genuchten equation, and five parameters for Fredlund and Xing equation. All the equation parameters are essentially determined by fitting. However, Brooks-Corey equation and van Genuchten equation cannot represent the SWCC for the entire range of matric suction. Although Fredlund and Xing equation was the best fitting equation among the SWCC equations, the method to determine the parameters was still complicated (see an example in Zhai and Rahardjo 2012). In this paper, a new equation is proposed based on the work by Liu et al. (2013). It is given as follows

$$\theta_w = \begin{cases} \theta_{w,\max} \text{ for } \theta_w > \theta_{w,\max} \\ \frac{a}{\psi}\left[1 + (\psi - 1)e^{-b\psi^c}\right] + d \text{ for } \theta_w \leq \theta_{w,\max} \end{cases} \tag{1}$$

In the equation, θ_w is the volumetric water content, and ψ is the suction with a unit of kPa. There are five parameters in total introduced in the equation. The three main parameters are *a*, *b*, and *c*, defining the curvature of a SWCC. Parameter *d* defines the asymptotic value of the curve. There is an additional parameter describing the maximum value of the volumetric water content, $\theta_{w,max}$.

Parameters *a*, *b*, and *c* are assumed to be positive. Mathematically, the valid range for variable ψ, which is dependent on *c*, is given as

$$\begin{cases} \psi \geq 0 \text{ for } c \geq 1 \\ \psi > 0 \text{ for } 0 < c < 1 \end{cases} \tag{2}$$

Parameter *c* is determined by trial and error. It is suggested that some systematic research on the parameter should be carried out for the convenience of engineering application. It can be seen that the new function has the following features. (1) The asymptotic line for the proposed function is θ_w = d. (2) The value of suction ψ varies linearly with the value of parameter *a*. This is clearly shown in Eq. (1). Therefore, parameter *a* is determined by the amplitude of the *S*-shape curve. (3) The limit value for θ_w, as ψ approaches zero, is as follows

$$\theta_w \longrightarrow \psi \to 0 \begin{cases} a \text{ for } c > 1 \\ a + ab \text{ for } c = 1 \\ \infty \text{ for } c < 1 \end{cases} \tag{3}$$

The function has a singular point at $\psi = 0$ for $c < 1$. The value of *a* can be determined from Eq. (3) for $c > 1$. For $c = 1$, an equation for parameter *a* and *b* is found, i.e., $a + ab = \theta_{w, \text{ for } x=0}$.

3 Simulations

In this section, the SWCCs for eight soils with nine cases are simulated via the proposed equation. The tests are reported in three papers. Values of equation parameters identified are listed in Tables 1, 2 and 3. As seen in Eq. (1), the determination of parameters *a*, *b* and

c needs special effort. In this study it is found that parameter c is relatively stable. As can be seen in the values listed in the Tables, it may be assumed as constant for a given soil. Parameter b is usually less than 0.5. It is a variable but usually its value changes gradually with influencing factors such as the value of net stress. Comparisons of equation simulations and experimental data are shown in Figs. 3, 4 and 5.

Table 1. Data for Khon Kaen loess

Tests and figures	θ_{max} (%)	a	b	c	d	Comments
Net stress = 200 kPa, Fig. 3	35.3	53	0.25	0.25	0	SWCCs flat initially
Net stress = 150 kPa, Fig. 3	29.2	78	0.4	0.25	0	

Table 2. Data for a sandy soil, a silty soil, and a clayey soil

Tests and figures	θ_{max} (%)	a	b	c	d	Comments
Sandy soil, Fig. 4	37.4	37.4	0.15	2	0	Influence of soil particle size clearly seen
Silty soil, Fig. 4	57.5	57.5	0.011	0.68	0	
Clayey soil, Fig. 4	91.3	91.3	0.017	0.4	0	

Table 3. Data for SWCCs reported by Mahmood et al. (2016)

Tests and figures	θ_{max} (%)	a	b	c	d	Comments
Soil 1, Fig. 5	41.4	28	0.26	0.5	5.9	For this set of data, $d \neq 0$
Soil 2, Fig. 5	47.3	33	0.2	0.5	8.2	
Soil 4, Fig. 5	31.4	20	0.15	0.5	4.9	
Soil 5, Fig. 5	52.1	36	0.16	0.5	9.5	

The data on Khon Kaen loess were reported by Likitlersuang and Sevilla 2013. The SWCCs for two net stresses are obtained, i.e., 150 and 200 kPa. Comparisons of simulations and data are shown in Fig. 3. The observed data are represented reasonably well by the proposed equation. The initial volumetric water content for the loess is essentially constant, i.e., $\theta_{w,max}$ = 29.2% for net stress being equal to 150 kPa, and $\theta_{w,max}$ = 35.3% for net stress being equal to 200 kPa. The maximum volumetric water content changes with net stress.

The experimental data reported by Fredlund and Xing (1994) are simulated. There are three types of soil: sandy soil, silty soil, and clayey soil. Comparisons of simulations and data for the three sets of tests are shown in Fig. 4. The influence of particle size on SWCCs is clearly seen in this set of test data. With the decrease in soil particle size, suction increases steadily, and the range of volumetric water content increases steadily. The experimental data of soil for both clay and sand are described satisfactorily, including different ranges in the variation for matric suctions and volumetric water contents.

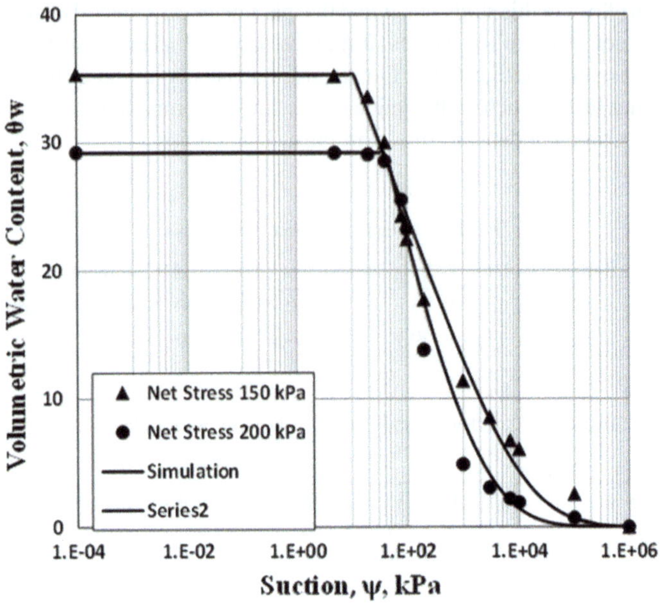

Fig. 3. SWCCs for Khon Kaen loess (data after Likitlersuang and Sevilla 2013)

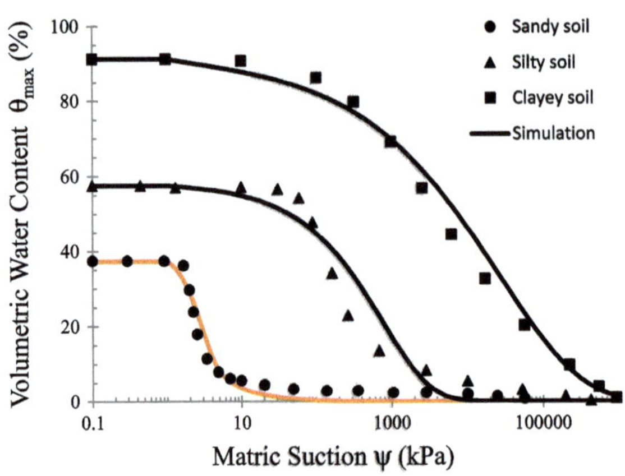

Fig. 4. SWCCs for a sand, a silty, and a clay (data after Fredlund and Xing 1994)

SWCCs for four types of soil reported by Mahmood et al. (2016) are also simulated. Comparisons of simulations and experimental data are shown in Fig. 5. The experimental data of these soils are modelled highly satisfactorily. It is also seen that the asymptotic value for this set of data is not zero.

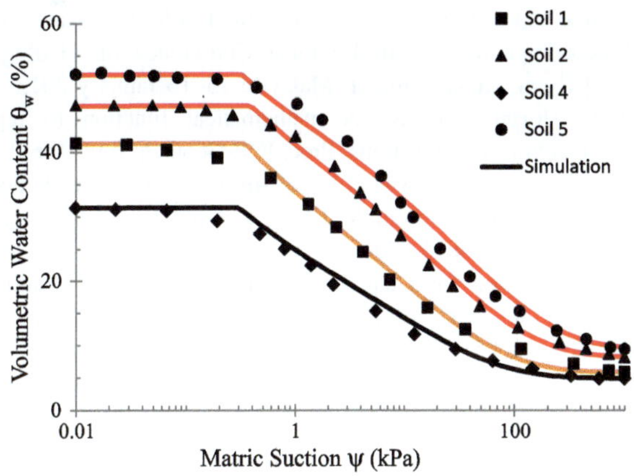

Fig. 5. SWCCs for soils reported by Mahmood et al. (2016)

4 Conclusions

In this paper, a new equation for SWCCs is proposed. The equation is an elemental function and thus can be employed conveniently in mathematical modelling and numerical analyses. There are five parameters in the equations. Two of the parameters, $\theta_{w,max}$ and d, can be determined from the SWCC straightforward, and the other three need to be determined by fitting. The SWCCs for eight soils with nine cases are simulated via the proposed equation, and it is shown that the proposed equation has the capacity to represent the curves of various soils satisfactorily.

Acknowledgements. This research was funded by the Ratchadapisek Sompoch Endowment Fund (2016), Chulalongkorn University (CU-59-055-CC).

References

Brooks, R.H., Corey, A.T.: Hydraulic properties of porous media. Colorado State University, Fort Collins, Colorado. Hydrology Paper No. 3 (1964, March)

Fredlund, D.G., Sheng, D., Zhao, J.: Estimation of soil suction from the soil-water characteristic curve. Can. Geotech. J. **48**, 186–198 (2011). https://doi.org/10.1139/t10-060

Fredlund, D.G., Pham, H.Q.: A volume-mass constitutive model for unsaturated soils in terms of two independent stress state variables. In: Miller, G.A., Zapata, C.E., Houston, S.L., Fredlund, D.G. (eds.) Proceedings of the 4th International Conference on Unsaturated Soils, ASCE, Carefree, Arizona, 2–6 April, vol. 1, pp. 105–134. American Society of Civil Engineers, Reston, VA (2006)

Fredlund, D.G., Xing, A.: Equations for the soil–water characteristic curve. Can. Geotech. J. **31** (4), 521–532 (1994). https://doi.org/10.1139/t94-061

Fredlund, D.G., Rahardjo, H., Fredlund, M.D.: Chapter 5 soil–water characteristic curves for unsaturated soils, unsaturated soil mechanics in engineering practice, John Wiley & Sons, Inc. (2012)

Likitlersuang, S., Sevilla, M.E.P.: Parametric study of models for soil–water characteristic curve of sands. In: Proceedings of the 5th Regional Conference on Geological Engineering, pp. 142–150. Citiel Hotel, Kuala Lumpur, Malaysia 15–16 January 2013

Liu, M.D., Xu, K.J., Horpibulsuk, S.: A mathematical function to represent S-shaped relationships for geotechnical applications. Proc. ICE-Geotech. Eng. **166**(3), 321–327 (2013)

Mahmood, K., Kim J.M., Ashraf, M.: The effect of soil type on matric suction and stability of unsaturated slope under uniform rainfall. KSCE J. Civil Eng. **20**(4), 1294–1299 (2016). https://doi.org/10.1007/s12205-015-0796-z

van Genuchten, M.T.: A closed form equation for predicting the hydraulic conductivity of unsaturated soils. Soil Sci. Soc. Am. J. **44**(5), 892–898 (1980). https://doi.org/10.2136/sssaj1980.03615995004400050002x

Zhai, Q., Rahardjo, H.: Determination of soil-water characteristic curve variables. Comput. Geotech. **42**, 37–43 (2012). https://doi.org/10.1016/j.compgeo.2011.11.010

Effects of Dry Density and Water Content on Mechanical Properties of Sand-Bentonite Buffer Material

Janaka J. Kumara(✉) and Takeshi Kodaka

Meijo University, Nagoya, Japan
janaka_eng@yahoo.com

Abstract. Mechanical properties of the buffer material of deep geological repository are influenced by various factors. The engineered barrier system of a deep geological repository is subjected to local groundwater flow after it is decommissioned. The changes in water content of buffer material could affect deformation and suction properties. In this study, the influence of water content and dry density on the mechanical properties of sand-bentonite buffer material were investigated. The triaxial compression tests were performed on sand-bentonite specimens of 1400, 1600 and 1800 kg/m^3 of dry density and 6, 12 and 18% of water content. The volumetric strains of specimens were evaluated using a newly built double-cell type triaxial testing apparatus. Total suction of specimens was measured using the chiller-mirror hygrometer technique. Total suction was measured on identical specimens prepared for suction measurement and triaxial compression tests. The results indicate that water content reduces deviator stress while dry density increases it. The results also suggest that water content changes strain-softening behaviour of relatively less saturated specimens into strain-hardening behaviour when water content is high. A high confining pressure (of 1.0 MPa) inclines towards strain-hardening behaviour than a small confining pressure (of 0.1 MPa). Water content also increases strain-hardening behaviour. In contrast, dry density reduces strain-hardening behaviour, particularly under a small confining pressure. The results also indicate that dry density reduces the magnitude of volumetric expansion. A high confining pressure encourages volumetric expansion. While water content decreases frictional behavior, it increases cohesive behaviour under a relatively low dry density (of 1400 and 1600 kg/m^3). In contrasts, water content reduces cohesive behaviour under a high dry density (of 1800 kg/m^3). Thus, a micro-scale analysis would produce more insights on this. While water content has huge influence on total suction, both dry density and confining pressure have no effects on it.

1 Introduction

Radioactive wastes are classified according to various standards. As illustrated in Fig. 1, the radioactive wastes can be classified into six different classes based on the half-life time of radioactive wastes (IAEA 2009). High level radioactive wastes need advanced disposal system to keep them away from people and environment as

M.-C. Weng et al. (eds.), *Current Geotechnical Engineering Aspects of Civil Infrastructures*, Sustainable Civil Infrastructures,
https://doi.org/10.1007/978-3-319-95750-0_9

radioactive wastes emit heat for very long time. It has been widely accepted that deep geological repository is a safe solution for high level radioactive wastes in many countries. The concepts and designs of deep geological repositories are still being discussed in many parts of the world.

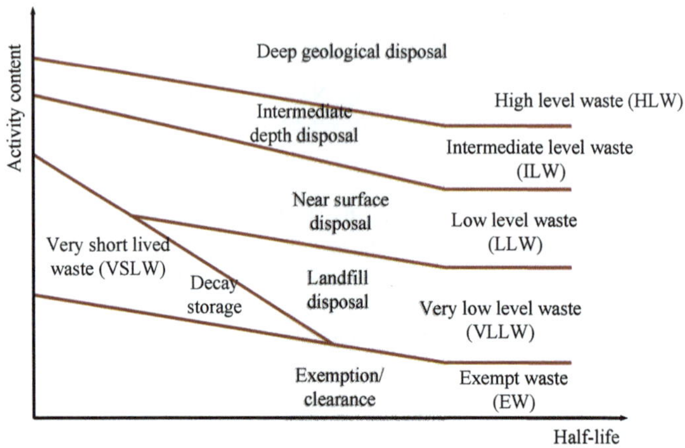

Fig. 1. A conceptual illustration of the radioactive waste classification (inspired by IAEA 2009)

The basic features of deep geological repository are shown in Fig. 2. As illustrated in Fig. 2, a deep geological repository is constructed in a deep stable hard rock. It could be several hundred meters below the ground surface. The nuclear wastes are stored in canisters either in horizontal drifts or vertical boreholes. As a part of the multi-barrier system, the buffer material is used to protect the radioactive wastes being infiltrated into the outside environment as illustrated in Fig. 2. The engineered barrier system is constructed by an expansive soil to allow swelling such that any voids can be filled by the buffer material. There have been various types of clayey soil discussed as potential buffer material in many countries. In Europe, MX-80 has widely been selected as a buffer material (Herbert and Moog 1999; Nakashima 2006; Hurel and Marmier 2010). In China, Gaomiaozi (GMZ) bentonite has recently been proposed as a potential buffer material (Ye et al. 2009). In Japan, Kunigel VI bentonite has been studied as a possible buffer material (Komine and Ogata 2004; Imbert and Villar 2006).

An engineered barrier provides various purposes including the mechanical stability for the waste canisters, serving as a buffer around it, sealing discontinuities in the boreholes and delaying water infiltration from the host rock (Ye et al. 2014). Bentonite-type materials are generally selected as a buffer material for an engineered barrier system because of its high swelling capacity, low permeability, micro-porous structure and good sorption properties (Pusch 1992; Komine and Ogata 1994). The buffer material during the construction and early stage of a deep geological repository remains unsaturated due to low water content. However, in its running time, water could infiltrate into the buffer material, and thus makes it saturated or partially-saturated. It could be several hundred or thousand years due to low permeability of the host rock. Swelling and permeability characteristics of bentonite materials have been studied in greater capacity compared to strength properties of bentonite-based buffer material (Cui et al. 2008; Schanz et al. 2010; Wang et al. 2012). In this study, we investigated the influence of water content and dry density on deformation and suction behaviour of sand-bentonite buffer material.

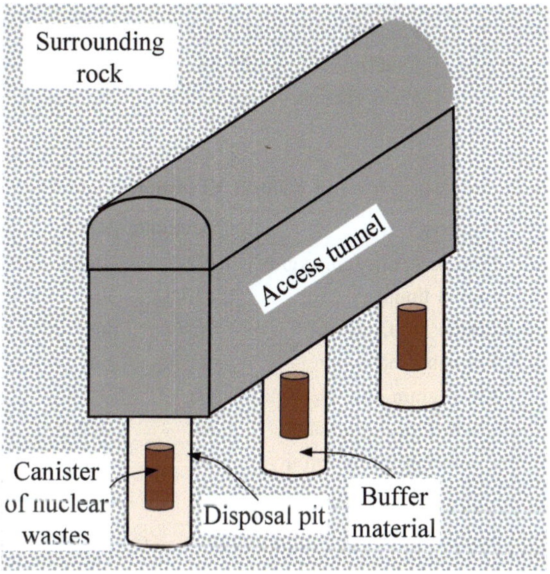

Fig. 2. A schematic diagram of a deep geological repository

2 Materials and Testing Program

The buffer material is prepared using 70% of bentonite and 30% of sand (by mass-based). This ratio has been accepted as a possible buffer material for the deep geological repositories in Japan (Mitachi 2008). The sodium-type bentonite contains around 48% of montmorillonite, which is an essential mineral for sealing properties. Kunigel VI also consists of quartz, which is an important mineral for thermal conductivity (Tang et al. 2008). The mineral compositions of Kunigel VI bentonite are given in Table 1. The index properties of bentonite and silica sand are given in Table 2. The grain size distribution of sand is shown in Fig. 3.

Table 1. Mineral compositions of Kunigel VI bentonite (Cui et al. 2008)

Mineral	Amount (%)
Montmorillonite	46–49
Pyrite	0.5–0.7
Calcite	2.1–2.6
Dolomite	2.0–2.8
Analcite	3.0–3.5
Feldspar	2.7–5.5
Quartz	29–38
Field organic	0.31–0.34

Table 2. Index properties of Kunigel VI bentonite and silica sand

Property	Bentonite	Sand
Particle density, ρ_s (kg/m^3)	2767	2666
Liquid limit, w_L (%)	430.5	n/a
Plastic limit, w_P (%)	26.7	n/a
Plasticity index, I_p	403.8	n/a
Mean grain size, D_{50} (mm)	n/a	0.230
Coefficient of curvature, C_c	n/a	0.865
Coefficient of uniformity, C_u	n/a	2.200

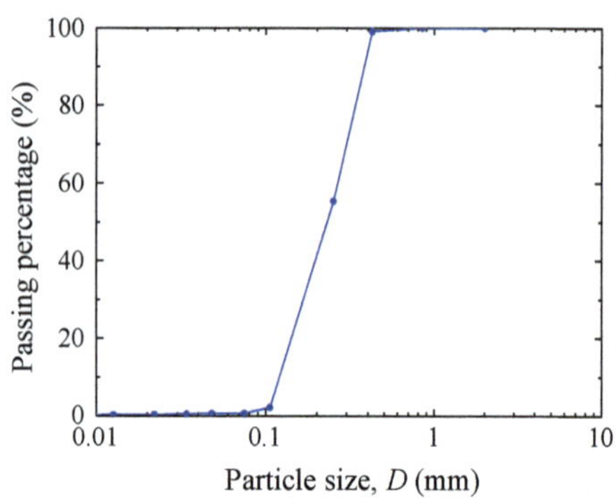

Fig. 3. Particle size distribution of sand

2.1 Triaxial Compression Test

The specimens for triaxial compression tests were prepared with 6, 12 and 18% of water content and 1400, 1600 and 1800 kg/m^3 of dry density. Two identical specimens of each condition were prepared for triaxial compression tests as the tests were performed under 0.1 and 1.0 MPa of confining pressure. First, pre-determined amount of sand and bentonite are mixed. Then, water is added to the sand-bentonite mixture using a high-pressurised sprayer, and simultaneously mixed water and the sand-bentonite mixture. As bentonite absorbs water and could make larger particles, water should be spread uniformly and mixed quickly. Then, the sand-bentonite mixture is put into the mold shown in Fig. 4a. The soil is placed in three layers, and each layer is given a hand compaction by a small tool shown in Fig. 4a. Then, it is compressed by a hydraulic jack shown in Fig. 4b. A prepared specimen is shown in Fig. 4c. Once the specimen reaches its pre-defined height (around 80 mm), it is removed from the mold by applying a hydraulic force. It is important to note that, height of the specimen again reduces slightly here. However, the reduction in height is affected by water content and dry density because the cohesion between the surfaces of the mold and the specimen is influenced by water content and dry density. Therefore, ± 2% difference in the design dry density and measured dry density is accepted for dry density. Finally, after trimmed both ends (roughly by 5 mm each), a specimen of 70 mm high and 35 mm diameter is prepared. A difference of ±5 is accepted for water content as it is measured by three disks of 20 g each (from roughly 170–220 g of the sand-bentonite mixture). Table 3 includes the details of the specimens prepared for triaxial compression tests. In Table 3, the measured values are reported.

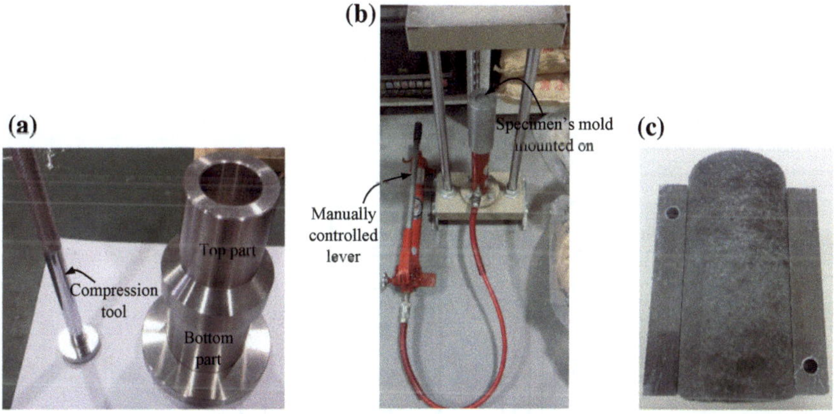

Fig. 4. The sample preparation: **a** the mold and its compaction tool, **b** hydraulic jack and **c** a prepared specimen

Triaxial compression test is performed using the newly built double-cell type triaxial testing apparatus shown in Fig. 5. The triaxial compression tests were performed under unconsolidated undrained condition. The specimens are covered with a 3 mm thick rubber membrane. The testing apparatus is designed such that the confining pressure applied to the outer cell also applies to the inner cell, in which the specimen is mounted. The volume change of specimens is measured by a burette attached to it as shown in Fig. 5. The shearing load is applied with a loading rate of 0.1%/min. The load is applied by a Mega-torque motor. The triaxial compression tests are performed based on the Japanese standards (JGS 2009).

Table 3. The details of the specimens for triaxial compression tests

Specimen notation	Water content, w (%)	Total density, ρ_t (kg/m^3)	Dry density, ρ_d (kg/m^3)	Degree of saturation, S_r (%)	Void ratio (e)
TT_6_1800_0.1	7.16	1925	1796	37.4	0.524
TT_6_1800_1.0	5.83	1900	1795	30.4	0.525
TT_12_1800_0.1	13.13	2029	1793	68.3	0.526
TT_12_1800_1.0	11.90	2021	1806	63.2	0.515
TT_18_1800_0.1	17.52	2084	1773	88.2	0.543
TT_18_1800_1.0	18.05	2101	1780	91.9	0.537
TT_6_1600_0.1	6.78	1681	1574	25.1	0.739
TT_6_1600_1.0	5.99	1688	1593	22.8	0.718
TT_12_1600_0.1	13.26	1778	1570	48.8	0.744
TT_12_1600_1.0	12.26	1791	1595	46.9	0.715
TT_18_1600_0.1	18.20	1889	1598	69.9	0.712
TT_18_1600_1.0	17.81	1894	1608	69.4	0.702
TT_6_1400_0.1	5.63	1460	1382	15.7	0.980
TT_6_1400_1.0	5.55	1473	1396	15.8	0.961
TT_12_1400_0.1	11.20	1554	1398	32.0	0.958
TT_12_1400_1.0	12.08	1562	1393	34.3	0.964
TT_18_1400_0.1	17.04	1689	1443	52.0	0.897
TT_18_1400_1.0	17.40	1603	1365	47.4	1.004

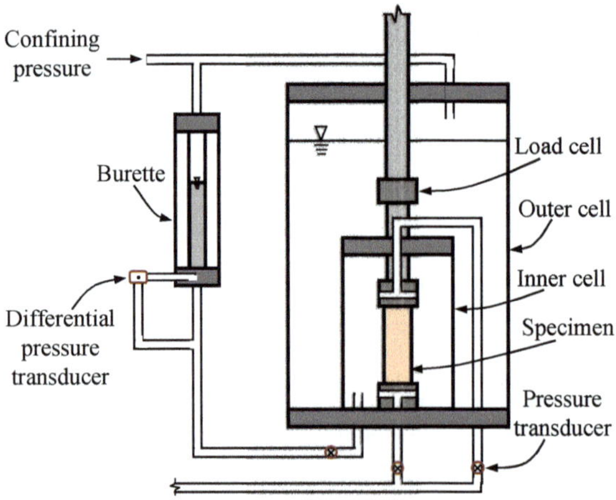

Fig. 5. A schematic diagram of the double-cell type triaxial testing apparatus

2.2 Total Suction

In addition to the specimens prepared for triaxial compression tests, identical specimens were also prepared for suction measurements. Then, a small piece around the half size of the sample cup is prepared from the specimen of 70 mm high and 35 mm diameter. When a sample is placed on the sample cup and inserts into the chamber, it starts to equilibrate with the headspace of the sealed chamber. Total suction is measured using the Dewpoint PotentiaMeter (WP4C) shown in Fig. 6. The device has a digital display to read the measurements. Total suction is measured using the chilled mirror hygrometer technique (Agus et al. 2010). The temperature of samples should be managed such that the temperature difference between a sample and the chamber should ideally be within 0 and −0.5 °C (it should always be negative). Suction measurements were also obtained on the specimens tested in triaxial compression tests.

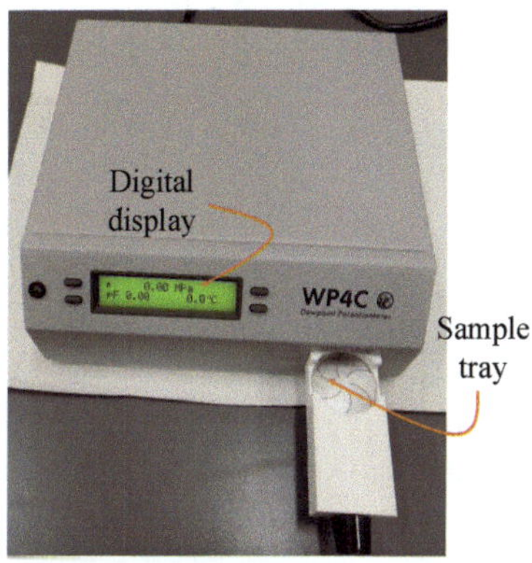

Fig. 6. The suction measurement device

3 Results and Discussion

Figures 7 and 8 show the influence of water content and dry density on stress-strain behaviour of sand-bentonite buffer material under a small (i.e., 0.1 MPa) and high confining pressure (i.e., 1.0 MPa) respectively. Figures 7 and 8 clearly indicate that water content decreases deviator stress while dry density increases it. Thus, it suggests that the buffer material of a deep geological repository reduces its strength after several years; could be several hundreds of thousands, when the local groundwater flow makes into the buffer material through the impermeable host rock. As indicated in Fig. 7a–c, water content also changes strain-softening behaviour of relatively dry specimen into strain-hardening behaviour under a small confining pressure regardless of dry density. Thus, the buffer material of a deep geological repository mainly yields strain-hardening behaviour in its long life time. Under a high confining pressure, as shown in Fig. 8a–c, many specimens yield strain-hardening behaviour compared to a small confining pressure. Therefore, the buffer material of a deep geological repository constructed in deep ground simply yields strain-hardening behaviour. We can see only the specimen of low water content (e.g., around 6%) and high dry density (i.e., 1800 kg/m^3) yields strain-softening behaviour under a high confining pressure. Thus, less likelihood of such behaviour in a deep geological repository in its long-term life cycle. The results also indicate that buffer materials of high water content under high confining pressure exhibit large deformations before shear failure as proven by strain-hardening behaviour in Figs. 7 and 8.

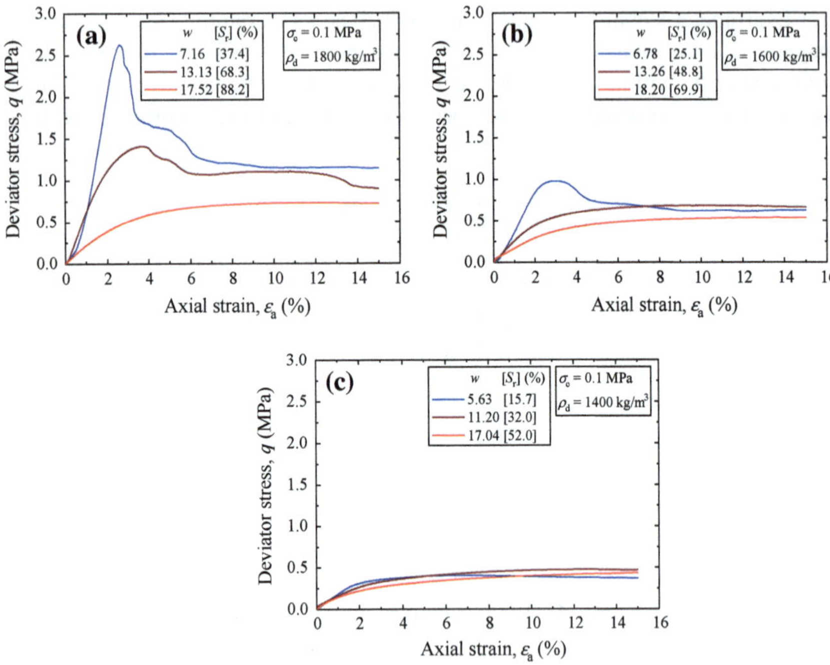

Fig. 7. The influence of water content and dry density on stress-strain behaviour under **a** 1800, **b** 1600 and **c** 1400 kg/m^3 of dry density and a small confining pressure, σ_c of 0.1 MPa (w is water content, S_r is degree of saturation and ρ_d is dry density)

Figures 9 and 10 show the influence of water content and dry density on the volumetric strain behaviour under a small and high confining pressure respectively. Under a small confining pressure (of 0.1 MPa), water content increases volumetric expansion as illustrated in Figs. 9a–c. In contrast, dry density discourages volumetric expansion. The results further indicate that a highly compacted specimen (e.g., 1800 kg/m^3 of dry density) encourages volumetric contractions, particularly until the specimens become fully saturated. When the specimens become full saturated, volumetric expansion appears as shown in Fig. 9a. Under a high confining pressure (of 1.0 MPa), the specimens yield only volumetric expansion as shown in Fig. 10a–c. It is also worth to note that unlike a small confining pressure (of 0.1 MPa), the highest water content (of 18%) does not yield the largest volumetric expansion as depicted in Fig. 10a–c. In fact, the specimens of around 12% water content yield the largest volumetric expansion under a high confining pressure. The results further indicate that dry density decreases the magnitude of volumetric expansion as indicated in Fig. 10a–c (except the sole specimen of 12% water content and 1800 kg/m^3 of dry density shown in Fig. 10a. This case needs to be further studied as it behaves differently to other specimens). Volumetric expansion is a key parameter in designing a deep geological repository as the buffer material should expand to occupy any voids made by chemical reactions of the radioactive wastes. A low dry density would increase volumetric expansion, but it would also reduce the strength as illustrated in Figs. 7 and 8. Therefore, the results suggest that dry density should be carefully designed along with the confining pressure (influenced mainly by the depth) and water content of the buffer material, which is mainly influenced by the initial water content of bentonite and permeability characteristics of the host rock.

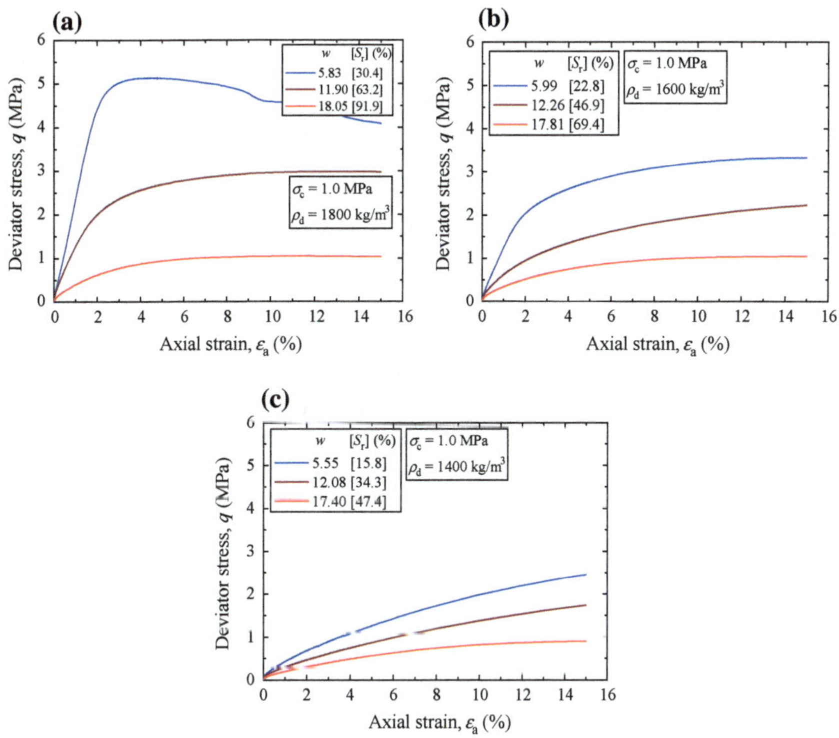

Fig. 8. The influence of water content and dry density on stress-strain behaviour under **a** 1800, **b** 1600 and **c** 1400 kg/m^3 of dry density and a high confining pressure, σ_c of 1.0 MPa (w is water content, S_r is degree of saturation and ρ_d is dry density)

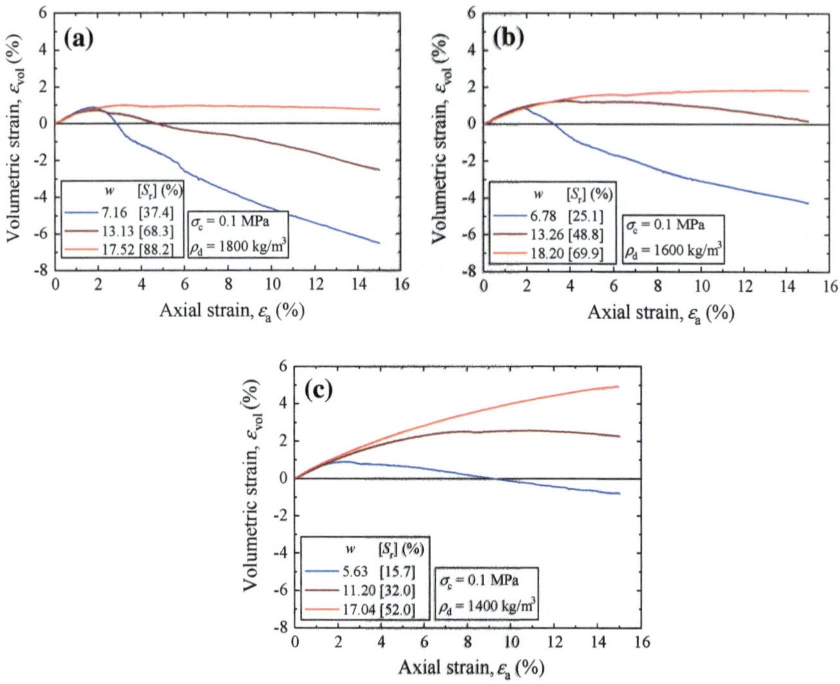

Fig. 9. The influence of water content and dry density on volumetric strain behaviour under **a** 1800, **b** 1600 and **c** 1400 kg/m^3 of dry density and a small confining pressure, σ_c of 0.1 MPa (w is water content, S_r is degree of saturation and ρ_d is dry density)

The strength parameters were evaluated using the Mohr Coulomb failure criteria. Figure 11 illustrates a typical Mohr stress circle. In Fig. 11, the data from an additional triaxial compression test under 0.5 MPa of confining pressure is also included to confirm the accuracy of the Mohr Coulomb failure criterion from two specimens. Figure 12a, b illustrate the variations of cohesion and internal friction angle with water content and dry density. Figure 12a depicts a highly compacted specimen (e.g., of 1800 kg/m^3 of dry density) reduces its cohesion with water content. In contrast, relatively less compacted specimens (e.g., of 1600 and 1400 kg/m^3 of dry density) increase cohesion with water content. A micro-level analysis would give more insights as water seems to produce bigger particles upon bentonite absorbing water. As shown in Fig. 12b, internal friction decreases with water content, and at highly wet specimens (e.g., water content of around 18%), there is no big variation due to dry density. In fact, both 1600 and 1800 kg/m^3 dry densities exhibit similar frictional angles under all water contents. Figure 13 illustrates the variation of compressive strength with water content and dry density. Except the specimens of loosely compacted (i.e., 1400 kg/m^3 of dry density) under a small confining pressure (of 0.1 MPa), all other specimens indicate both water content and dry density reduce the compressive strength. The loosely compacted specimen under a small confining pressure seems to slightly increase compressive strength with water content up to 12%, and then slightly reduces. In Fig. 13, the ultimate strength in strain-softening specimens and peak strength in strain-hardening specimens are considered as the compressive strength.

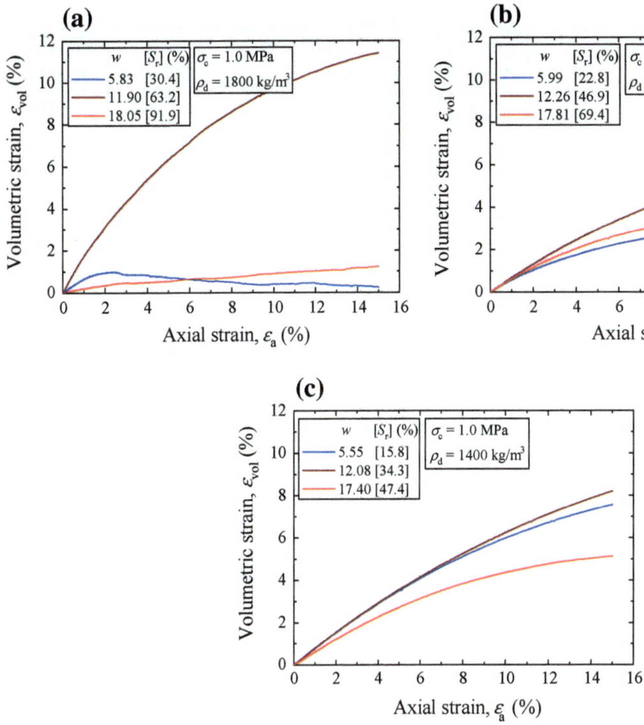

Fig. 10. The influence of water content and dry density on volumetric strain behaviour under **a** 1800, **b** 1600 and **c** 1400 kg/m^3 of dry density and a high confining pressure, σ_c of 1.0 MPa (w is water content, S_r is degree of saturation and ρ_d is dry density)

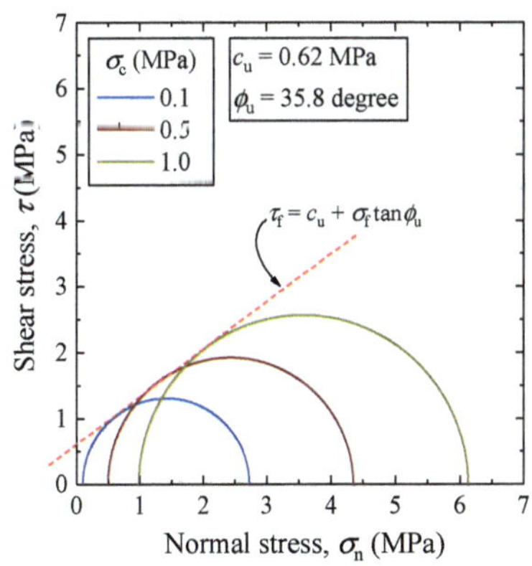

Fig. 11. A typical Mohr circle and its failure criteria (of 6% of water content and 1800 kg/m^3 of dry density)

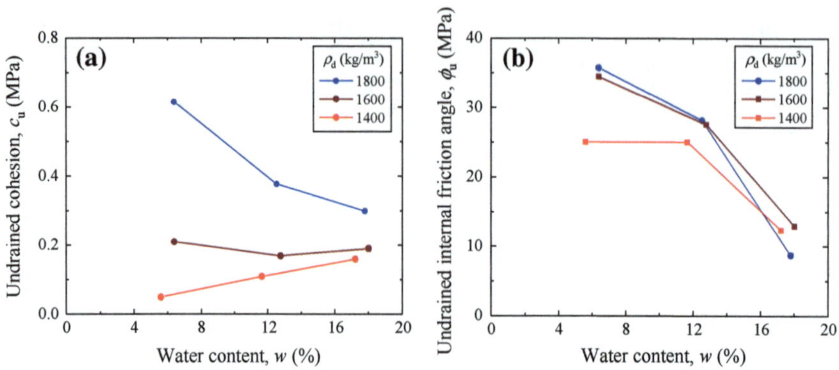

Fig. 12. The influence of water content and dry density, ρ_d on **a** cohesion and **b** internal friction angle

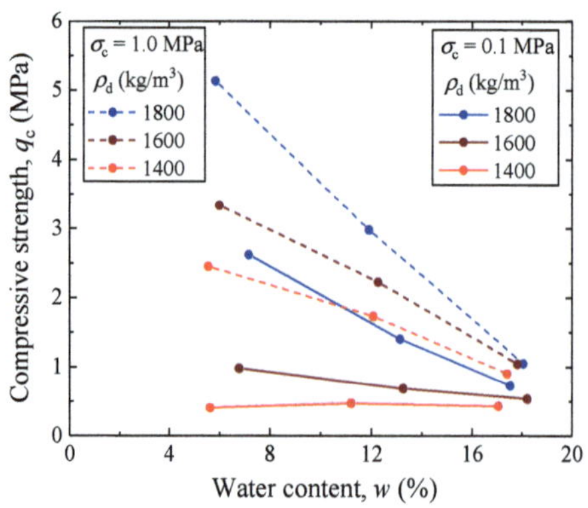

Fig. 13. The influence of water content and dry density on compressive strength

Figure 14 shows the variation of total suction with water content and dry density. The results clearly indicate that total suction drastically decreases with water content. However, dry density does not affect total suction. Also, it indicates that there is no effects of confining pressure on total suction or any effects from shear loading from triaxial compression tests. This is an important observation as it hints that suction of a buffer material is not affected by the depth (i.e., confining pressure) or loading. Thus, an earthquake might not change suction properties of the buffer material of a deep geological repository. However, more studies are needed before concrete conclusions are drawn on this.

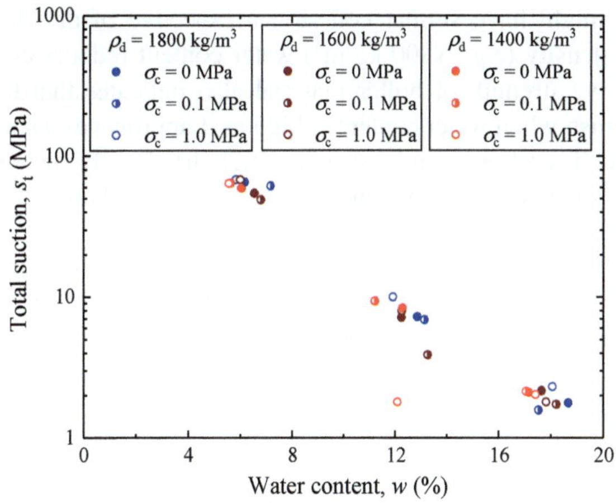

Fig. 14. The influence of water content and dry density, ρ_d on total suction (σ_c is confining pressure; 0.1 and 1.0 of it means the s_t measurements after triaxial tests)

4 Conclusions

In this study, triaxial compression tests and suction measurements were performed on sand-bentonite buffer material of deep geological repository. The buffer material was prepared with 6, 12 and 18% of water content and 1400, 1600 and 1800 kg/m^3 of dry density. The following conclusions are drawn from the study.

Water content decrease deviator stress while dry density increases it under both a small (0.1 MPa) and high confining pressure (1.0 MPa). The results also indicate that the specimens of less water content and high dry density under a small confining pressure yield strain-softening behaviour. In contrast, highly compacted specimens of high water content yield strain-hardening behaviour. Under a high confining pressure, all the specimens except very low water content and high dry density yield strain-hardening behaviour. In contrast, highly compacted specimen of less water content yields strain-softening behaviour under a high confining pressure.

Under a small confining pressure, water content encourages volumetric expansion. Thus, the specimens of high water content yield volumetric expansion whereas the specimens of low water content exhibit volumetric contraction. Under a high confining pressure, all specimens yield volumetric expansion. The results also indicate that irrespective of confining pressure, dry density reduces the magnitude of volumetric expansion.

Water content reduces the internal friction angle while dry density has less effects on internal friction angle. In fact, irrespective of dry density, the specimens of high water content (e.g., around 18%) tend to produce similar internal friction angle. Therefore, it would suggest dry density has less effects on frictional behaviour of a buffer material after several thousand years. In contrast, except high dry density (e.g., 1800 kg/m^3), water content increases cohesion. Therefore, it suggests after a long time, a buffer material might have similar cohesion irrespective of dry density. However, under a high dry density (e.g., 1800 kg/m^3) water content reduces cohesion.

The compressive strength of buffer material also indicates that the effects of dry density on it diminish when water content is high as it approaches a similar value under high water content. Therefore, it suggests that a high dry density such as 1800 kg/m^3 is not necessary to maintain a high compressive strength when a buffer material is infiltrated by local groundwater flow.

The results of total suction indicates that while water content drastically reduces total suction, dry density has no effects on total suction. The results also suggest that confining pressure has no effects on total suction, which would indicate that an earthquake might have no effects on suction behaviour. However, it needs further studies to draw concrete conclusions on this.

Acknowledgements. The research funding comes from the JSPS Postdoctoral Fellowship (P17074), which is greatly appreciated. Assoc. Prof. Ying Cui of Yokohama National University, Japan and Mr. Keisuke Takeuchi, a former graduate student of Meijo University are also acknowledged for their supports during laboratory experiments.

References

Agus, S.S., Schanz, T., Fredlund, D.G.: Measurements of suction versus water content for bentonite-sand mixtures. Can. Geotech. J. **47**(5), 583–594 (2010). https://doi.org/10.1139/T09-120

Cui, Y.J., Tang, A.M., Loiseau, C., Delage, P.: Determining the unsaturated hydraulic conductivity of a compacted sand-bentonite mixture under constant-volume and free-swell conditions. Phys. Chem. Earth **33**(S1), 462–471 (2008). https://doi.org/10.1016/j.pce.2008.10.017

Herbert, H.J., Moog, H.C.: Cation exchange, interlayer spacing, and water content of MX-80 bentonite in high molar saline solutions. Eng. Geol. **54**(1–2), 55–65 (1999). https://doi.org/10.1016/S0013-7952(99)00061-7

Hurel, C., Marmier, N.: Sorption of europium on a MX-80 bentonite sample: experimental and modelling results. J. Radioanal. Nucl. Chem. **284**(1), 225–230 (2010). https://doi.org/10.1007/s10967-010-0476-x

IAEA: Classification of Radioactive Waste—General Safety Guide. International Atomic Energy Agent (IAEA), Vienna (2009)

Imbert, C., Villar, M.V.: Hydro-mechanical response of a bentonite pellets/powder mixture upon infiltration. Appl. Clay Sci. **32**(3–4), 197–209 (2006). https://doi.org/10.1016/j.clay.2006.01.005

JGS: Method for Unconsolidated-Undrained Triaxial Compression Test on Soils (JGS 0521). Japanese Geotechnical Society Standard, Tokyo (2009)

Komine, H., Ogata, N.: Experimental study on swelling characteristics of compacted bentonites. Can. Geotech. J. **31**(4), 478–490 (1994). https://doi.org/10.1139/t94-057

Komine, H., Ogata, N.: Predicting swelling characteristics of bentonite. J. Geotech. Geoenviron. Eng. **130**(8), 818–829 (2004). https://doi.org/10.1061/(ASCE)1090-0241

Mitachi, T.: Mechanical behavior of bentonite-sand mixtures as buffer materials. Soils Found. **48**(3), 363–374 (2008). https://doi.org/10.3208/sandf.48.363

Nakashima, Y.: H_2O self-diffusion coefficient of water-rich MX-80 bentonite gels. Clay Miner. **41**(2), 659–668 (2006). https://doi.org/10.1180/0009855064120211

Pusch, R.: Use of bentonite for isolation of radioactive waste products. Clay Miner. **27**(3), 353–361 (1992)

Schanz, T., Arifin, Y.F., Khan, M.I., Agus, S.S.: Time effects on total suction of bentonites. Soils Found. **50**(2), 195–202 (2010). https://doi.org/10.3208/sandf.50.195

Tang, A.M., Cui, Y.J., Le, T.T.: A study on the thermal conductivity of compacted bentonites. Appl. Clay Sci. **41**(3–4), 181–189 (2008). https://doi.org/10.1016/j.clay.2007.11.001

Wang, Q., Tang, A.M., Cui, Y.-J., Gelage, P., Gatmiri, B.: Experimental study on the swelling behaviour of bentonite/claystone mixture. Eng. Geol. **124**, 59–66 (2012). https://doi.org/10.1016/j.enggeo.2011

Ye, W.M., Cui, Y.J., Qian, L.X., Chen, B.: An experimental study of the water transfer through confined compacted GMZ bentonite. Eng. Geol. **108**(3–4), 169–176 (2009). https://doi.org/10.1016/j.enggeo.2009.08.003

Ye, W.M., Borrell, N.C., Zhu, J.Y., Chen, B., Chen, Y.B.: Advances on the investigation of the hydraulic behavior of compacted GMZ bentonite. Eng. Geol. **169**, 41–49 (2014). https://doi.org/10.1016/j.enggeo.2013.11.003

Geotechnical Investigation on Compacted Clay Liner Blended with Reactive Material

Polapala Sai Pradeep[1], Mayakrishnan Muthukumar[1(✉)], and Sanjay Kumar Shukla[2]

[1] Department of Structural and Geotechnical Engineering, School of Civil and Chemical Engineering, VIT University, Vellore, India
mmuthukumar@vit.ac.in

[2] Discipline of Civil and Environmental Engineering, School of Engineering, Edith Cowan University, Perth, Australia
s.shukla@ecu.edu.au

Abstract. Clays are used for the construction of waste containment barriers, mainly because of its low hydraulic conductivity. In general, at several places, it may not be possible to obtain the clay satisfying the essential requirements as a liner material. Hence at many places commercially available bentonites is often used for the construction of clay liners. Bentonite undergoes significant changes in volume upon saturation. Several physical and chemical stabilization methods are in practice to stabilize the bentonite. Previous studies report that contaminants with high concentrations attack the clay minerals leading to an increase in their permeability. Exfoliated vermiculite, finds several applications in construction industry, and it is also found to be good absorbent material. Hence this study has been carried out to find the optimal percentage of vermiculite that can be added to the bentonite so that it satisfies the essential requirements of clay liners and it can also absorb the heavy metals from the contaminants. The tests for index properties and engineering properties of bentonite-vermiculite mixtures were performed. The test results show that 30% of vermiculite can be added to bentonite, so that the essential requirements of the liner can be maintained.

1 Introduction

Bentonite, a commercially available clay, is commonly used in the construction of hydraulic barriers because of its very low hydraulic conductivity. Bentonite, composed of montmorillonite clay mineral, undergoes significant changes in volume because of which it fails to control the migration of leachate present in the dumped wastes. In order to minimize the volume change attributes several stabilization techniques are adopted, most popular being sand-bentonite mixtures. In sand-bentonite mixtures, sand is added to bentonite to a certain percentage so that the essential requirements for use as landfill liners are maintained. These requirements include permeability equals to or less than 10^{-7} cm/s, finer fraction greater than 30%, plasticity index greater than 10% and cation exchange capacity (CEC) greater than 10 meq/100 g (Daniel 1993; Rowe et al. 2004). Extensive research has been carried out on sand-bentonite mixtures and many

M.-C. Weng et al. (eds.), *Current Geotechnical Engineering Aspects of Civil Infrastructures*, Sustainable Civil Infrastructures,
https://doi.org/10.1007/978-3-319-95750-0_10

correlations are available with several geotechnical properties (Stern and Shackelford 1998). Al-Rawas et al. (2006) reported that bentonite mixed with 20% sand showed very low hydraulic conductivity but the hydraulic conductivity of sand-bentonite mixture increased drastically under the permeation of chemical solutions. However due to desiccation, the bentonite present in the liner shrinks resulting in an increase in the hydraulic conductivity of the liner (Mukherjee and Mishra 2017). Several studies evaluated the use of various additives such as lime, cement, fly ash, etc. to control desiccation cracking. Initial results of these studies indicated that soil shrinkage reduced, but in many cases the additives resulted in an increased hydraulic conductivity with a decrease in soil plasticity.

Vermiculite, an interesting clay mineral, found in many parts of the world, belongs to 2:1 phyllosilicate mineral, composed of hydrated magnesium-aluminium-iron sheet silicates containing water molecules within the layered structure (Mitchell and Saga 2005). This vermiculite when heated between 870 and 1100 °C, the internal structure gets disturbed and the volume expands to several times that of the original material. This process is known as exfoliation (Marcos and Rodriguez 2014). This exfoliated vermiculite becomes light in weight under the exfoliation process with good thermo-insulating properties. Exfoliated vermiculite is used in making light-weight concrete (Silva et al. 2010), thermal insulations (Abidi et al. 2015) and also extensively used as a good adsorbent material instead of activated carbons in the removal of heavy metals because of its high CEC and reactive surfaces. Vermiculite as an absorbent has been studied for the removal of Mercury (Do Nascimento and Masini 2014), Chromium (Sis and Uysal 2014; Dultz et al. 2012), Zinc (Sis and Uysal 2014), Cadmium (Mathialagan and Veeraraghavan 2003) and even dyes from textile industries (Mysore et al. 2005; Duman et al. 2015; Da Fonseca et al. 2006, Marcos and Rodriguez 2014). As the exfoliated vermiculite is found to be an effective and excellent material used for the removal of heavy metals from the waste disposal, this paper is aimed at investigating the potential use of exfoliated vermiculite-bentonite mixtures instead of sand-bentonite mixtures as a landfill liner, so that it can absorb the heavy metals in the leachate as well as it can stabilize the volume change attributes of bentonite. The main objective of the study was to investigate the potential use of bentonite vermiculite mixtures. The present study focuses on permeability characteristics as well as swelling and shrinkage characteristics of bentonite with the addition of various percentages of exfoliated vermiculite. One of the main objective is to determine the optimum percentage of bentonite vermiculite mixture so that the essential requirements of compacted clay liner can be maintained.

2 Experimental Investigation

2.1 Test Materials

Commercially available sodium bentonite was used in the present study. Various properties of the bentonite have been determined and reported in Table 1. Exfoliated vermiculite of particle size in the range of 1–4 mm was used in the present study. The various physical and chemical properties of the vermiculite supplied by the manufacturer are reported in Table 2. The SEM analysis has also been carried out on exfoliated vermiculite sample. Figure 1 shows the SEM image of exfoliated vermiculite. Cation Exchange Capacity was determined for both bentonite and vermiculite using methyl blue test. The CEC is found to be 14 meq/100 g for bentonite and 16 meq/100 g for vermiculite. The CEC of the vermiculite is more when compared to bentonite, so addition of vermiculite will not affect the essential requirement of vermiculite.

Table 1. Properties of the bentonite used for this study

Soil properties	Value
Specific gravity	2.73
Liquid limit (%)	129
Plastic limit (%)	30
Shrinkage limit (%)	1.8
Dry density	1.64 g/cc
Optimum moisture content	20%
Free swell index (FSI) (%)	128

Table 2. Physical and chemical properties of the exfoliated vermiculite

Physical properties	
Bulk density	64–160 kg/m^3
Water holding capacity	220–325% by wt.
MOH	1–2
pH	6–9
Chemical properties	
Silica	39.4%
Magnesium	25.2%
Alumina	8.8%
Potassium	4.5%
Iron	4.0%
Calcium	1.8%
Carbonate	1.4%
Titanium	0.8%
Fluorine	0.5%

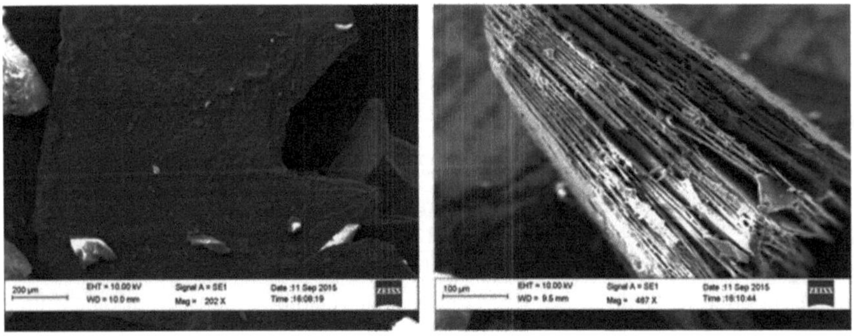

Fig. 1. SEM image of exfoliated vermiculite

2.2 Index Properties Tests

Tests for index properties such as liquid limit, plastic limit, shrinkage limit and shrinkage strain tests have been conducted on bentonite-vermiculite blends as per the ASTM standards. Liquid limit and Plastic limit tests were conducted as per ASTM D 4318-10 (1996). Shrinkage limit test was determined according to ASTM D427-04 (2007). After determining shrinkage limit, diameter and thickness of the soil pat were measured using digital calliper to determine radial and axial shrinkage strains. The axial and radial shrinkage strains were calculated using the following expressions.

Axial shrinkage strain

$$\epsilon_a = \frac{h_i - h_f}{h_i} \times 100 \tag{1}$$

where: h_i is the initial thickness, and h_f is the final thickness.

Radial shrinkage strain

$$\epsilon_r = \frac{D_i - D_f}{D_i} \times 100 \tag{2}$$

where: D_i is the initial diameter, and D_f is the final diameter.

2.3 Shrinkage Limit Tests

Standard Proctor compaction, hydraulic conductivity and swelling tests have been conducted on bentonite and bentonite mixed with various percentages of vermiculite as discussed earlier. Initially Standard Proctor compaction tests have been conducted as per ASTM D 698-12 (1996) for determining the optimum moisture content for the bentonite blended with different percentages of vermiculite. After performing compaction tests, hydraulic conductivity tests were performed for the bentonite and bentonite-vermiculite blends compacted at their respective optimum moisture content and maximum dry density obtained from compaction tests. Hydraulic conductivity tests were performed using rigid wall permeameter according to ASTM D5856 (1996). Soil was compacted in the mould and the mould was placed in the water bath for four days to allow for full saturation of the sample. The permeability tests were performed using de-aired tap water. Swelling tests were conducted in the prefabricated mould of diameter 10 cm and 15 cm height. Soil was compacted in the mould in three layers of thickness of 2.5 cm. A dial gauge was fitted at the top of the soil layer. After setting the dial gauge reading to zero, water was added continuously at the top of the clay bed and the swelling was monitoring continuously at various time intervals until the swelling attained equilibrium.

3 Discussion of Test Results

Figure 2 shows the variation of liquid limit and plastic limit for varying vermiculite content. It is observed that the liquid limit of the bentonite reduced marginally when compared to plastic limit. The plastic limit tends to increase with percentage of vermiculite. Upon addition of 30% vermiculite to bentonite the liquid limit was found to be 120%, whereas the plastic limit found to be 62%, resulted in a plasticity index of 58%. Figure 3 shows the variation of shrinkage limit of bentonite blended with varying percentage of vermiculite. The shrinkage limit of the bentonite is 1.8%, which is considered to undergo large volume changes (Holtz and Gibbs 1956). Upon addition of vermiculite, the shrinkage limit tends to increase rapidly. The shrinkage limit of bentonite blended with 20% vermiculite is 17.9% and beyond 20% vermiculite there is no significant increase in shrinkage limit. According to Holtz (1956), if the shrinkage limit is greater than 16%, the soil does not undergo significant volume change. Addition of vermiculite also reduces the shrinkage strains of bentonite (Fig. 4). Both axial and radial shrinkage tends to decrease with increase in vermiculite content. Axial strain reduced to 47.3% and radial shrinkage reduced to 39.4%, when 30% vermiculite content was blended with bentonite.

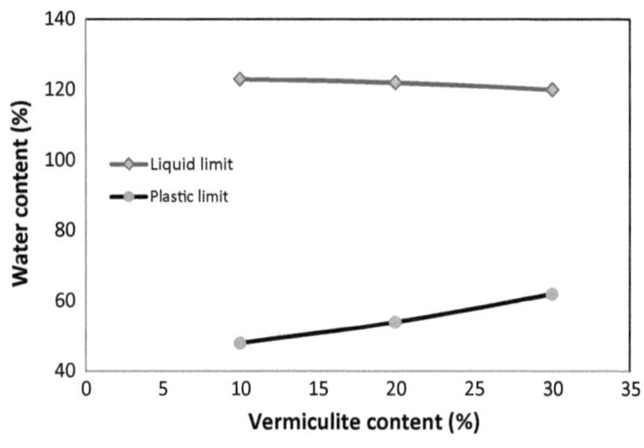

Fig. 2. Variation of liquid limit and plastic limit

Figure 5 shows the variation of maximum dry density and optimum moisture content (OMC) of bentonite with different percentage of exfoliated vermiculite contents. The reduction in dry unit weight of bentonite with an increase in vermiculite content is due to the replacement of soil with vermiculite of lower specific gravity leading to increase in volume and decrease in unit weight. Conversely, the OMC increased with higher vermiculite content. The increase in OMC with increased vermiculite content was likely due to the larger amount of water absorbed by the vermiculite because of its inherent water retention capacity.

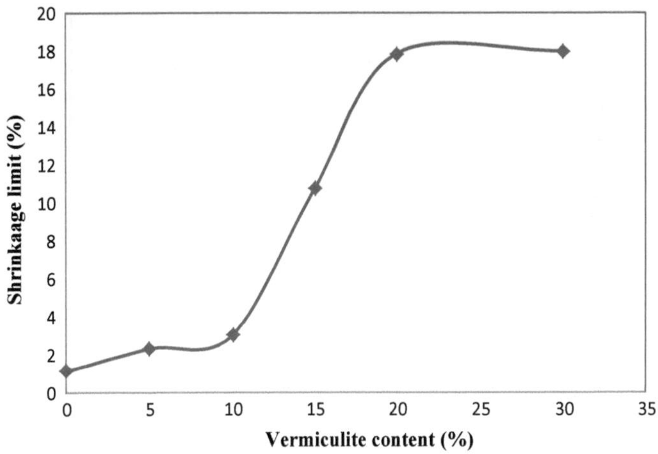

Fig. 3. Variation of shrinkage limit with varying vermiculite content

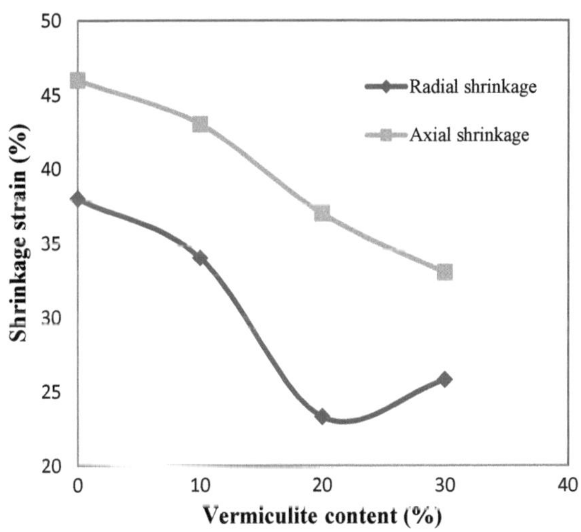

Fig. 4. Variation of shrinkage strains

Achieving specific value of permeability is one of the important criteria for the clay to be used as a liner material. Permeability of the clay should be less than 1×10^{-7} cm/s. Variation in the hydraulic conductivity of vermiculite-bentonite mixtures is reported in Fig. 6. For bentonite sample, the hydraulic conductivity was found to be 3.86×10^{-09} cm/s. The hydraulic conductivity reduced to 7.10×10^{-9} cm/s, when vermiculite content was 10%, beyond 10% vermiculite content, the permeability tends

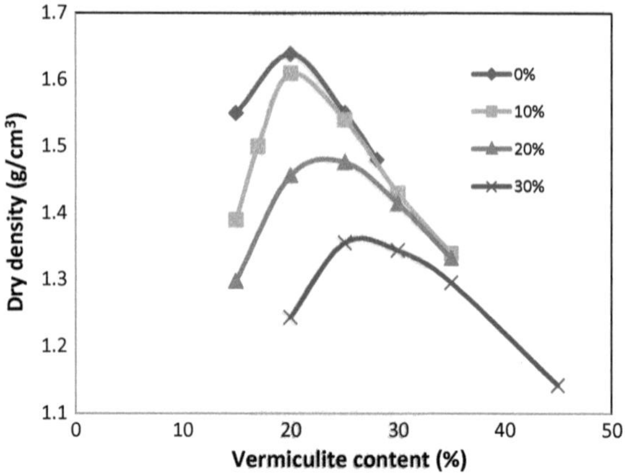

Fig. 5. Compaction characteristics of bentonite-vermiculite blends

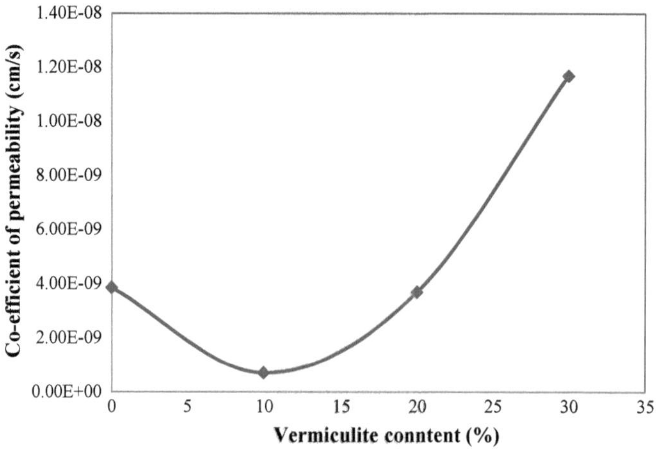

Fig. 6. Variation of coefficient of permeability for varying vermiculite content

to increase. For bentonite blended with 30% vermiculite, the permeability was found to be 1.93×10^{-07} cm/s. Beyond 30% vermiculite, the permeability was found to be high, which is greater than the essential requirement of the clay liner. Swelling of bentonite was also reduced due to the addition of vermiculite (Fig. 7). Swelling decreased from 14.1 to 6.9 mm when vermiculite was varied from 0 to 30%.

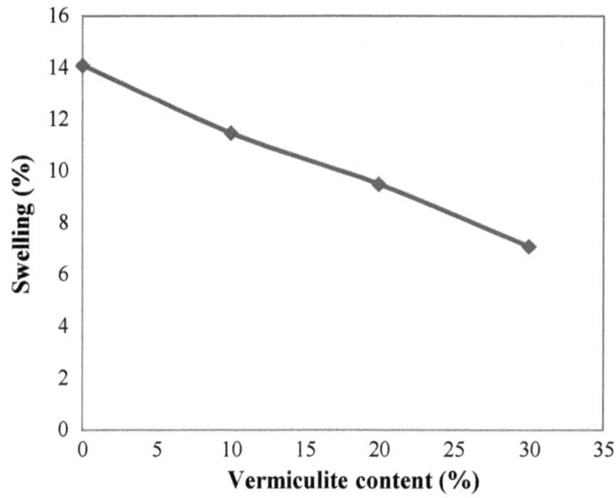

Fig. 7. Variation of swelling

4 Conclusions

Based on the results and discussion presented here, the following conclusions can be made.

With addition of 30% vermiculite to bentonite, the liquid limit is found to be 120% and the plastic limit found to be 62%; resulting in a plasticity index of 58%. The shrinkage limit tends to increase rapidly. The shrinkage limit of bentonite blended with 20% vermiculite is 17.9% and beyond 20% vermiculite, there is no significant increase in shrinkage limit.

Axial strain reduces to 47.3% and radial shrinkage reduces to 39.4%, when 30% vermiculite content is blended with bentonite. Swelling also tends to decrease with an increase in vermiculite content. The swelling reduces to 51%, when 30% of vermiculite content is added to bentonite.

The hydraulic conductivity reduces to 7.10×10^{-9} cm/s, when vermiculite content is 10%, beyond 10% vermiculite content, the permeability tends to increase. For bentonite blended with 30% vermiculite, the permeability is found to be 1.93×10^{-07} cm/s. Beyond 30% vermiculite the permeability is found to be high, which is greater than the essential requirements of the clay liner.

References

Abidi, S., Nait-Ali, B., Joliff, Y., Favotto, C.: Impact of perlite, vermiculite and cement on the thermal conductivity of a plaster composite material: experimental and numerical approaches. Compos. B Eng. **68**, 392–400 (2015)

Al-Rawas, A.A., et al.: Sand-attapulgite clay mixtures as a landfill liner. Geotech. Geol. Eng. **24** (5), 1365–1383 (2006)

ASTM D 4318-10: Standard Test Methods for Liquid Limit, Plastic Limit, and Plasticity Index of Soils. ASTM International, West Conshohocken (1996)

ASTM D 698-12: Standard Test Method for Laboratory Compaction Characteristics of Soil Using Standard Effort. ASTM International, West Conshohocken (1996)
Da Fonseca, M.G., Wanderley, A.F., Sousa, K., Arakaki, L.N.H., Espinola, J.G.P.: Interaction of aliphatic diamines with vermiculite in aqueous solution. Appl. Clay Sci. **32**, 94–98 (2006)
Daniel, D.E., Shan, H.Y., Anderson, J.D.: Effects of partial wetting on the performance of the bentonite component of a geosynthetic clay liner. Geosynthetics' 93. Industrial Fabrics ASSA. Int. St. Paul Minn. **32**, 1483–1496 (1993)
Do Nascimento, F.H., Masini, J.C.: Influence of humic acid on adsorption of Hg(II) by vermiculite. J. Environ. Manag. **143**, 1–7 (2014)
Dultz, S., An, J.H., Riebe, B.: Organic cation exchanged montmorillonite and vermiculite as adsorbents for Cr(VI): effect of layer charge on adsorption properties. Appl. Clay Sci. **67–68**, 125–133 (2012)
Duman, O., Tunç, S., Polat, T.G.: Determination of adsorptive properties of expanded vermiculite for the removal of C. I. Basic Red 9 from aqueous solution: kinetic, isotherm and thermodynamic studies. Appl. Clay Sci. **109–110**, 22–32 (2015)
Holtz, W.G., Gibbs, H.J.: Engineering properties of expansive clays. Trans. ASCE **121**, 641–677 (1956)
Marcos, C., Rodríguez, I.: Applied Clay Science Some effects of trivalent chromium exchange of thermo-exfoliated commercial vermiculite. Appl. Clay Sci. **90**, 96–100 (2014)
Mathialagan, T., Vararaghavan, : Adsorption of cadmium from aqueous solution by vermiculite. Sep. Sci. Techol. **38**(2003), 57 (2003)
Mitchell, J.K., Soga, K.: Fundamentals of soil behavior. Wiley, Hoboken, NJ (2005)
Mukherjee, K., Mishra, A.K.: Performance enhancement of sand-bentonite mixture due to addition of fiber and geosynthetic clay liner. Int. J. Geotech. Eng. **11**(2), 107–113 (2017)
Mysore, D., Viraraghavan, T., Jin, Y.C.: Treatment of oily waters using vermiculite. numerical approaches. Compos. B Eng. **68**, 392–400 (2005)
Rowe, J.K., Quigley, R.M., Brachman, R.W.I., Booker, J.R.: Barrier Systems for Waste Disposal Facilities, vol. 95, 2nd edn, pp. 1–8. Spon Press, London (2004)
Silva, L., Ribeiro, R., Labrincha, J., Ferreira, V.: Role of lightweight fillers on the properties of a mixed-binder mortar. Cement Concr. Compos. **32**(1), 19–24 (2010)
Sis, H., Uysal, T.: Removal of heavy metal ions from aqueous medium using vermiculite. J. Environ. Manag. **143**, 1–7 (2014)
Stern, R.T., Shackelford, C.D.: Permeation of sand-processed clay mixtures with calcium chloride solutions. J. Geotech. Geoenviron. Eng. **124**(3), 231–241 (1998)

Author Index

A
Afshani, Alireza, 1
Awwad, Talal, 13

B
Banerjee, Lalima, 69
Bhandari, Gupinath, 69

C
Chawla, Sowmiya, 69, 83

D
Datta, Manoj, 24

G
Gruzin, Vladimir, 13
Gupta, Randhir Kumar, 83

H
Hasan, Grant, 1
Hirokazu, Akagi, 1

I
Inti, Sundeep, 38
Isenhower, William M., 54

K
Kim, Vladimir, 13
Kodaka, Takeshi, 109
Kumara, Janaka J., 109

L
Likitlersuang, Suched, 101
Liu, Martin D., 101

M
Muthukumar, Mayakrishnan, 124

N
Nath, Brajendra, 83

P
Pant, Aali, 24
Pradeep, M. Polapala Sai, 124

R
Ramana, Gunturi Venkata, 24

S
Sharma, Megha, 38
Shukla, Sanjay Kumar, 124

T
Tandon, Vivek, 38

V
Vasquez, Luis G., 54

X
Xu, Daqing, 54

M.-C. Weng et al. (eds.), *Current Geotechnical Engineering Aspects of Civil Infrastructures*, Sustainable Civil Infrastructures,
https://doi.org/10.1007/978-3-319-95750-0

MIX
Papier aus verantwortungsvollen Quellen
Paper from responsible sources
FSC® C105338

If you have any concerns about our products,
you can contact us on
ProductSafety@springernature.com

In case Publisher is established outside the EU,
the EU authorized representative is:
Springer Nature Customer Service Center GmbH
Europaplatz 3, 69115 Heidelberg, Germany

Printed by Libri Plureos GmbH
in Hamburg, Germany